Forms of Becoming

Forms of Becoming

THE EVOLUTIONARY BIOLOGY
OF DEVELOPMENT

Alessandro Minelli

Mark Epstein, translator

PRINCETON UNIVERSITY PRESS   PRINCETON AND OXFORD

First published in Italy under the title *Forme del divenire*. Copyright © 2007 Giulio Einaudi editore s.p.a., Torino

English translation © 2009 by Princeton University Press

Requests for permission to reproduce material from this work should be sent to Permissions, Princeton University Press

Published by Princeton University Press, 41 William Street, Princeton, New Jersey 08540
In the United Kingdom: Princeton University Press, 6 Oxford Street, Woodstock, Oxfordshire OX20 1TW

Library of Congress Cataloging-in-Publication Data

Minelli, Alessandro.
   [Forme del divenire English]
   Forms of becoming : the evolutionary biology of development / Alessandro Minelli ; Mark Epstein, translator.
      p.   cm.
   Includes bibliographical references and index.
   ISBN 978-0-691-13568-7 (hardcover : alk. paper)   1. Developmental biology.
2. Evolution (Biology) I. Title.
   QH491.M559513 2009
   571.8—dc22   2008028825

British Library Cataloging-in-Publication Data is available.

The translation of this work has been funded by SEPS
Segretariato Europeo per le Pubblicazioni Scientifiche

Via Val d'Aposa 7 - 40123 Bologna - Italy
seps@seps.it - www.seps.it

This book has been composed in Minion

Printed on acid-free paper. ∞
press.princeton.edu

Printed in the United States of America

10 9 8 7 6 5 4 3 2 1

Contents

PART THREE: Origins

Preface

Form and Function is the title of a classic book published in 1916, in which Edward Stuart Russell reconstructed the history of animal morphology from Aristotle to the early 1900s—a long period of time in which the most important and revolutionary event in the history of biology took place, including the advent of a world incorporating evolutionary change. And yet for Russell the most radical choice between alternative views in the study of animal form was not pre-Darwinian conceptions versus those dominated by the methods and priorities of evolutionary biology. It was instead the more ancient and perhaps never fully resolved opposition which, at the beginning of the nineteenth century, saw the two giants of comparative anatomy, Georges Cuvier and Etienne Geoffroy Saint-Hilaire, facing one another. The first championed a view that Russell called teleological and which consists in claiming the primacy of function over form, while the second defended, with equal conviction and authority, the view that Russell called morphological, which claims the primacy of form over function.

And today? Which are the most important questions concerning the multiplicity of forms of living organisms? Much water has of course since passed under the bridges, and not only those on the Seine, whose banks are a few feet from the Muséum d'Histoire Naturelle where, for almost forty years, Cuvier and Geoffroy Saint-Hilaire worked side by side—and it has certainly not invalidated the dichotomy that set the two French men of science against each other two hundred years ago. Something naturally has changed since then. Today we are researching both the mechanisms by means of which organisms are constructed, and the dialectical relationships they develop with the forever mutable environment in which they live. In other words the understanding of living forms today involves two quite distinct branches of biology: developmental and evolutionary.

In regard to the first branch what matters is form. The only functional aspects that deserve attention are those that pertain to the mechanisms that are the basis for the construction of forms, without regard for the achievements that these will be able to realize in either the short or the long term. Developmental biology also accommodates hopeless "monsters," such as calves with two heads or fruit flies with four wings, insofar as they are forms to which it is possible to give life.

For evolutionary biology, however, what matters is survival and reproduction. This presupposes an efficient use of the resources that the environment offers and, therefore, an appropriate level of efficiency on the part of the animals' organs. The forms that exist (or, better, those that survive the process of natural selection) satisfy the functional criteria established by the environmental milieu.

Paradoxically, therefore, it is precisely in evolutionary biology that primary attention to the function of organs survives. Cuvier, consigned to us by history as the staunch defender of fixed characteristics, was the fiercest proponent of this cause. A science like developmental biology, on the other hand, for which only the modest time-scale in which the individual is constructed seems relevant, even today champions the primacy of form over function central to Geoffroy Saint-Hilaire's vision. While it would not be historically accurate to attribute truly evolutionary ideas to Geoffroy Saint-Hilaire, he certainly was far removed from sharing his rival's firm belief in the unchanging nature of species.

It is also true that few researchers in evolutionary biology today read the works of Cuvier to find inspiration for their own work. Similarly it would be difficult for researchers in developmental biology to find inspiration in the works of Geoffroy Saint-Hilaire, although, as we shall soon see, the latter statement needs some further clarification.

For Cuvier, comparing the anatomical structure of a cat, a sparrow, and a lizard was a legitimate (and interesting) activity, since the three animals have many organs and apparatuses in common. On the other hand, the comparison of a cat to a butterfly, or a sparrow to an oyster, would have been pointless, given the great distance separating their respective organizational designs. Today, however, it is possible to attempt such daring comparisons, above all because modern developmental genetics has shown that many important stages in the construction of an animal take place, in species as different as an insect and a mammal, under the direction of the same genes. An extremely interesting discovery, but one that gives rise to new and more difficult questions.

The greatest problem at this point is no longer understanding what animals as different as a mouse and a fruit fly have in common, but, on the contrary, finding the causes of their diversity. And it is not sufficient, either, to hand off the problem to evolutionary biology, only to be faced with the reply that the mouse and the fruit fly are different because in the course of generations their ancestors have had to confront different environmental conditions, which, little by little, have selected the two different animal forms which we see today. The fact is that the problem of the differences between different living things remains also, at least in part, a problem of developmental biology. A problem made more complex by the fact that nature seems incapable of producing many forms that, in theory, would seem to represent only very modest variations compared to other forms that, instead, are actually produced. Development seems to have its obligatory points of passage, in which natural selection has no way of intervening. It is limited to choosing between the different variants that are available to it.

At this point we have to ask ourselves if it will ever be possible to reach an adequate comprehension of living forms and their evolution, as long as we remain tied to the traditional separation between developmental and evolutionary biology. The negative replies to this question are becoming increasingly insistent. In response to this, a discipline is taking shape that aims at integrating the concepts, the problem sets, and the methods of inquiry that pertain to the two traditions, and it is called evolutionary developmental biology, or evo-devo.

It is in terms of evo-devo that I present an interpretation of living forms in these pages. I should perhaps more correctly state: of animal forms. I shall write of plants and other living things only very briefly, due not only to my own background in zoology, but also to the fact that to date evolutionary developmental biology has concerned itself only marginally with plants, and not at all with fungi. There is, however, a treasure in the pages of the scientific literature devoted to botanical subjects and it is waiting to be re-read and re-interpreted from an evo-devo perspective. If the reading of these pages results in the recruitment to the new discipline of even only a couple of people interested in plants and fungi, this small book will have already fulfilled its purpose.

The book is divided into four sections. In the first I introduce some fundamental concepts of the comparative method, in the context of the history of biology. I then proceed to illustrate the unequal distribution of animal forms in the hypothetical space of expected forms. We shall thus see that the already existing forms are clustered in several privileged areas, leaving some unsuspected voids corresponding to forms that, for some reason, do not exist in nature. In the second section I discuss the limits of the gene's role in producing the forms of living organisms, with a critique of the current concept of the genetic program. And finally I clarify how to understand today's notion of "development of organisms." In the third section I invite the reader to explore some consequences of an interpretation of living organisms in terms of an evolutionary developmental biology—a choice that obliges us to assume a flexible, perhaps a pluralist, attitude toward the many traditional concepts of comparative morphology—from the abstract (what is homology?) to those that refer to real objects (what is a larva?). The fourth section, with the epilogue that concludes the book, is devoted to origins. The origin of legs, for instance, or the origin of the subdivision of the body into segments, such as we observe in an earthworm or a millipede. More generally, it is devoted to the origin of evolutionary innovation, the final link between the problems of developmental biology, which must tell us how it is possible to build these forms, and evolutionary biology, which must tell us how they changed over the course of time.

A few sincere and necessary thanks conclude these introductory pages. To do justice to all the people who sparked my interest in the topics treated in this book, I would need an extremely long list that would include scholars I have never met except through their writings. I will limit myself, therefore, to only three names that represent three generations. First, Pietro Omodeo, who since the time I wrote my doctoral dissertation under his guidance, made me appreciate both the value of theoretical reflection in biology and the importance (and fascination) of a frequent revisiting of past authors. Then Wallace Arthur, the first of the biologists of my generation with whom, starting at the end of the 1980s, I could engage in a dialogue on the subject of evolutionary developmental biology. And finally Giuseppe Fusco, the first of my students to courageously accompany me on this new and fascinating adventure.

Sincere thanks, also to Michele Luzzatto, for his constant encouragement, during these last few years, to write (and conclude!) this small book. To him, as to Lucio Bonato and Giuseppe Fusco, I also owe many precious comments on a first draft of the book.

For this U.S. edition I am deeply indebted to Robert Kirk, who was first responsible for getting my book included in the editorial program of Princeton University Press and afterward followed it carefully through the production process; and to Mark Epstein, who in a very accurate and sensible way selected the English words through which my thoughts are being conveyed in these pages.

PART ONE

Forms and Numbers

Chapter 1

Unity in Diversity

———————◆◆◆———————

Two Skeletons

Pierre Belon du Mans, doctor, naturalist, and traveler, is one of the most significant figures in the zoology of the sixteenth century. His was an adventurous life, brought to a tragic end by an assassin in the Bois de Boulogne, Paris, in April of 1564 or, perhaps, 1565. His most famous works are a natural history of fish and a natural history of birds. They are written in the colloquial tongue (French) and not in Latin, the sign of a spirit that wanted to be free of the shackles of tradition in order to suggest an original, innovative reading of natural events.

Even today it is worthwhile to leaf through Belon's ornithological work, if for no other reason than to peruse pages 40 and 41, almost entirely taken up with the figures of two skeletons: on the left that of a man, and on the right of a bird. Even this simple juxtaposition may seem unusual and perhaps irreverent. Man and bird are shown by Belon on a plane of perfect equivalence, postmortem, which at the time was probably a component of the collective imagination reminiscent of the *danses macabres* of penitential iconography—an equivalence reinforced by the ring attached to the skull, which ideally allows one to suspend each skeleton from a hook, so as to aid examination (fig. 1).

This explicit and almost brutal presentation of a human skeleton as an object worthy of study for natural history is itself a historical choice: in particular because this is not a medical textbook. And Belon's drawings assume a further, extraordinary value precisely because they encourage the reader to perform a careful work of

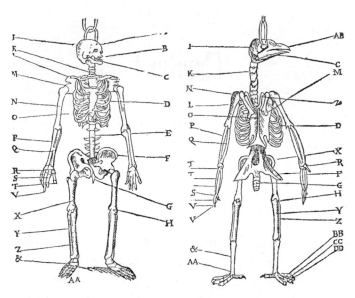

Figure 1. Comparison between the skeletons of a human being and a bird.
From Pierre Belon du Mans, *L'histoire de la nature des oyseaux*, Paris 1555.

comparison. In fact the two skeletons' bones are marked by a letter or a couple of letters each, and the same letters are utilized each time to indicate which bones, in Belon's opinion, are the same in the human being and the bird.

With this illustrated guide, the examination could be undertaken, arbitrarily, in either direction. If one was more familiar with the human skeleton, one would use it as a guide to help one understand the bird's skeleton. If instead one was more familiar with the latter, one would utilize it as a model for the human skeleton. There is moreover no doubt that in the comparison offered by the *Histoire de la nature des oyseaux* the author took for granted the reader's greater familiarity with the human skeleton, as contrasted to that of the bird. Belon's attitude was perhaps influenced by the medical studies he had undertaken, even though he obtained his doctoral degree at a fairly advanced age. Moreover the theme of the book was the natural history of birds, and therefore every notion borrowed from other living organisms, including human beings, could only have an instrumental value. But one should not interpret the fact that a human skeleton was used as a model for the study of the bird's skeleton as a survival of the old saying according to which man is the measure of all things. Man and bird, at the skeletal level at least, were considered equivalent by Belon.

The reader will begin to wonder what these two woodprints of 1555 have to do with the problems that today concern a newly formed discipline like evolutionary developmental biology. Please bear with me and accompany me in the following pages on a path that includes important milestones that have marked the history of the comparative method in biology. We shall soon arrive at the heart of those problems that most interest us.

Clashes at the Muséum

Belon having died, we can skip two centuries and a half, and remain in France, where, among the smoking ruins of the ancien régime, and Bonaparte's rising star, a center of scientific research has been born that will soon offer the international community some of the most innovative ideas in the realm of the natural sciences.

It had been Louis Daubenton, previously a trusted advisor to Georges-Louis Leclerq de Buffon, who had suggested to deputy Lakanal the guiding criteria the Convention would adopt when instituting the Muséum d'Histoire Naturelle on June 10, 1793. From its founding the Muséum had been conceived as a research center, a repository of precious research materials, and a place dedicated to public education. A group of *professeurs* was in fact called to supervise the activities of the different departments. The chair in vertebrate zoology was assigned to Etienne Geoffroy Saint-Hilaire, whom we shall meet several times in this book, while the position of *professeur* of the invertebrates was assigned, following orders from above, to Jean-Baptiste Monet de Lamarck, who as a consequence had to abandon his favorite botanical studies (in 1778 he had already published a *Flore françoise*). At this juncture, however, we are not interested in talking about Lamarck, but rather about Geoffroy Saint-Hilaire and Cuvier, three years Geoffroy's senior, whom Geoffroy requested be appointed to the professorship in comparative anatomy. Over the course of the following years, the two zoologists experienced moments of sincere friendship and fruitful professional collaboration, but also moments of intense rivalry and animosity.

Cuvier, in his work *Le règne animal*, proposed a subdivision of animal species into four large groups or *embranchements*: the vertebrates, the articulates, the mollusks, and the radiates. According to Cuvier these large groups corresponded to four organizational designs, so different from one another that any comparison between them was an arbitrary exercise. The situation within each *embranchement*, however, is very different: because its species are constructed according to a largely shared structural design, it is legitimate, in any of these species, to search for anatomical parts corresponding to those of other species, precisely as in the comparison Belon had proposed between the skeleton of the human being and of the bird.

But Geoffroy could not believe that nature would follow entirely different designs to generate those that, ultimately, are none other than different species belonging to the one and only animal kingdom. In other words he was convinced that all animal species share a substantially common structural design. Certainly this unity of design is more easily discernible when we compare two similar animal species: for instance two species that have a spinal column, or a trunk that is articulated into segmented units and supported by various pairs of appendages, which in their turn are also made up of interconnected articulated segments. According to Geoffroy, however, even the barriers that Cuvier established between the different *embranchements* are not absolute. It is simply necessary to be more ingenious in overcoming them, finding criteria of comparison that are applicable to animals as different as a crayfish and a fish.

The problem in this latter case is not only due to the fact that the crayfish has an exterior skeleton, with all its muscles on the inside, while the fish has an internal skeleton, with the muscles around it. A no less serious problem is, in comparative terms, the position of the main axis of the nervous system. In fish, and vertebrates in general, there is a spinal cord that runs along the back of the animal, protected by the vertebrae. In the crayfish, and in arthropods in general, as well as in many other animals, we instead find a ventral gangliar chain below the intestine. Not to worry, Geoffroy observed. In both cases there is a longitudinal nervous axis and we should not be concerned if in the vertebrates it is dorsal, while in arthropods it is ventral. In the end, it is still reasonable to think that the nervous system axis is always basically the same even if it occupies (or seems to occupy) different positions. Instead, what is the basis for our affirming that what we call "back" in a fish is "the same thing" as that which we call "back" in a crayfish? If only we could hypothesize that what we call the *ventral* aspect of a crayfish is equivalent to the *dorsal* aspect of a fish, the presumed contrast between the two anatomical arrangements would be reduced to fairly minor differences.

Nevertheless, it would still have been easy to refute Geoffroy's reasoning by maintaining that his supposed "solution" could be reduced to semantics, unless the opposite was true, and in reality it was the traditional anatomical descriptions that were hostage to arbitrary lexical choices, such as those on which Cuvier's position depended. From this point of view, the problem will not be reopened until the last years of the twentieth century. We shall discuss the issue again shortly.

In the meantime, however, in the cultural atmosphere of a Paris where Cuvier's authority had been progressively increasing, due to his undisputed scientific accomplishments and his ties to political power, Geoffroy's position, which suggested ways to compare vertebrates and Cuvier's articulates, did not have much hope of establishing itself. And the situation dramatically worsened when two of Geoffroy's students proposed a new and much less daring comparison, its two terms now being represented by a vertebrate and a mollusk.

Squid and Vertebrate

Meyranx and Laurencet examined what we know as cephalopods—the cuttlefish, the squid, and the octopus—as representative of the *embranchement* of the mollusks. And there are good reasons to attempt a comparison between vertebrates and cephalopods: more specifically the presence in both of particularly complex and efficient eyes, and of a brain capable of performances attained only with difficulty by most invertebrates. It is important to note, though, that the resemblances between fish and squid end here. The squid for instance has neither a skull nor a spinal column, but only a very thin internal shell (the "pen," even lighter and more fragile than the better known "bone" of the cuttlefish). And the general arrangement of the internal organs in the two animals is also different, because in the squid, as in other cephalopods, the digestive tube, which is relatively short, is folded into a *U* shape, the anal opening relatively close to the mouth. But it is precisely this arrangement of the organs that particularly concerned Geoffroy's two students, by showing that the "simple" bending of the main body axis of a vertebrate, folding the animal in on itself, greatly reduced the difference between the structural design of cephalopods and that of vertebrates.

This new formal application of the comparative method, suggested by the conviction that all animals share one common basic anatomical design, was too much for Cuvier. He was probably mostly fearful that the extension of comparative exercises between representatives of his different *embranchements* would favor the diffusion of "transformistic" ideas, such as those that Lamarck had defended in his works and above all in the *Philosophie Zoologique* of 1808.

In any case the debate between Cuvier and Geoffroy, which flared up in 1830 at the Académie des Sciences, was heated and continued for several sessions, between February and May of that year. A comment by Goethe, who was in Paris at the time, and who, writing to a correspondent, hinted at a volatile situation that was about to explode in the French capital, remains famous. At that time, Paris confronted one of its many revolutionary episodes, more precisely the one that would lead to the abdication of Charles X. But it was not to the barricades on the streets that Goethe alluded in his letter, but rather to the confrontation between the two great zoologists in the rooms of the Académie des Sciences.

Time would in large measure vindicate Geoffroy's positions, in regard both to the general principle of the unity of a structural design common to all animals, and to more specific questions, such as the equivalence of the dorsal aspect of vertebrates and the ventral aspect of arthropods and other animals. In fact, in the context of evolutionary developmental biology, Geoffroy Saint-Hilaire has, in recent years, come to impersonate the role of precursor to this new discipline on several occasions.

Topsy-turvy

Geoffroy, as we have suggested, "resolved" the problem by drawing the crayfish with its belly in the air next to a vertebrate viewed from the back and seen from above. Perhaps, he suggested, the distinction between belly and back is not really as certain or as profoundly rooted in animal organization as one would think. Ultimately, this distinction is above all dictated by the manner in which the animal relates to the external environment. The belly is the aspect facing the substrate, the sole on which the snake and the snail glide, the side from which the four legs of a gazelle or a crocodile reach toward the earth. And the back is, quite simply, the opposite aspect, the furthest from the substrate, that which an observer can see if looking down from above unless a shell or a protective carapace is interposed. The situation is less clear in the case of an animal that walks on two legs, like a human being, but even here the comparison with more conventional quadrupeds is not without merit. The distinction between belly and back becomes more arbitrary in the case of the earthworm, which, in its existence as a digger, is always completely surrounded by the substrate, almost as if it were an "all-belly" animal. There is moreover no zoology book that expresses any doubts as to the dorsoventral polarity of the earthworm, or that draws a heterodox transversal section. An established point of reference is the central nervous cord, the double ventral gangliar chain. But why ventral?

One must note that in many cases the distinction between belly and back cannot be inferred from an obvious structural polarity determined by the manner in which the animal moves around the world. The snail for instance glides on a flat sole that we have no difficulty in designating as ventral, but in cases like the earthworm the distinction derives instead from our comparative work with other animals in which the polarity is more obvious. The fact remains, however, that in the fruit fly and in the crayfish the gangliar nervous chain is on the side closest to the substrate, while the spinal cord of the cat and the snake are instead on the opposite side.

But is there not any more objective fashion to recognize an animal's belly and its back? Perhaps it is possible that in animals as different as arthropods and vertebrates the distinction between belly and back is controlled by the same genes, just as the different positions along the anteroposterior axis of very diverse sets of animals respond to a common molecular code that can be traced back to the expression of the *Hox* genes, a family of genes that we shall discuss later.

Fifteen years ago, a brief note published in the prestigious British journal *Nature* announced that it was once more time to give Geoffroy Saint-Hilaire his due for the daring comparisons he had undertaken. In fact in vertebrates and insects, the first stages in the differentiation of the longitudinal nervous cords are controlled by the same pair of genes, those which in the fruit fly have been designated *short gastrulation* and *decapentaplegic* and which in vertebrates have their precise equivalent in, respectively, the *chordin* and *Bone Morphogenetic Protein-4* genes. These genes are responsible for distinguishing, inside the most external layer of the embryo (the ectoderm), between the cells that will remain to form the animal's integument and those that will form its nerves and brain. There are therefore good reasons to affirm the equivalence between the ventral gangliar chain of the fruit fly and the dorsal nervous cord of vertebrates.

Chapter 2

Archetypes

The Primeval Plant

Goethe's interest in natural history is well known. Those who have read the *Voyage to Italy* will remember his visit to the botanical gardens of Padua and the inspiration that seems to have come to him on that occasion from observing the leaves of a plant: a plant living today that everyone now knows as "Goethe's palm." On a later trip to the botanical gardens in Palermo, in the course of the same voyage, the idea of the great plasticity of the leaf took root in his mind. The leaf, a primal element that, profoundly transformed, is perceptible in the elements that constitute the flower: not only in the sepals, which are usually green like a typical leaf, but also in the petals, which clearly differ from the sepals in both color and texture. In this interpretation, even the most internal parts of the flower, those with a reproductive function—on the one hand the stamens, on which the pollen matures, on the other the carpels, which taken together constitute the ovary—are seen as modified leaves.

Later, moving to the comparative anatomy of vertebrates, Goethe identifies the vertebra as another primal element capable of transformations, and he then postulates an equivalence between vertebrae and cranial bones. This is occurring at the beginning of the nineteenth century, when natural history, above all in Germany, shows the strong influence of the *Naturphilosophie* of Friedrich Wilhelm Joseph Schelling, who attributes an absolute and infinite character to nature and describes it as a vital principal that animates all organisms—moved by an internal finality and guided by an internal dialectic of attractive and repulsive forces. The borders between natural history and the philosophy of nature sometimes become uncertain and problematic, as in many pages by Lorenz Oken. In those same years (more precisely, in 1815), one of Goethe's countrymen, Johann Baptist von Spix, devotes a long essay to the topic of cephalogenesis, the origin of the head (in vertebrates): in this work one finds ideas very similar to Goethe's.

At around the same time, Etienne Geoffroy Saint-Hilaire proclaims the fundamental unity of the organizational design of all animals and undertakes a previously mentioned program of daring comparisons. Geoffroy's exercise consists of the tracing of correspondences between the structural aspects of different animals, without however imagining an ideal common design, of which all existing species would only be alternate versions. An ideal design emerges, instead, repeatedly, in the German zoological and botanical literature. An authentic icon of this attitude is the figure of the "primitive plant" (*Urpflanze*) drawn by Pierre Jean François Turpin and published for the first time in 1837, in a French translation of Goethe's morphological works. This ideal plant is as distant as one can imagine from our notion of a primal organism, which a century and a half of evolutionary biology has made it easy for us to conceive (fig. 2).

To be clear: it is not that everything runs without a hitch in the imagination of a contemporary evolutionary biologist. It is too easy, in fact, to slide into the "ancient = simple" equivalence, an identification that carries with it a trace, perhaps unconscious, of the preevolutionary notion of the *scala naturae,* a notion according to which the creation is oriented, following an ideal line of succession, to form a sequence that originates with inanimate creatures and then proceeds, first to the level of plants, then to the "inferior" and "superior" animals, up to human beings, and, beyond human beings, to angels and archangels, and finally ends with the Supreme Creator. In one of the following chapters, we will have the opportunity to observe that the direction of evolution has often been toward simpler, and not more complex forms. In any case to imagine that plants with flowers derive from terrestrial plants devoid of flowers, and that these in turn, at a greater remove, can be traced back to the simple forms of

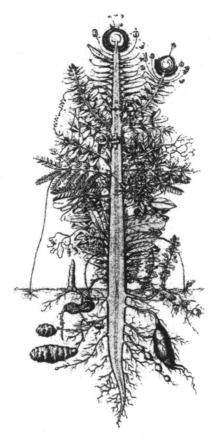

Figure 2. Turpin's *Urpflanze*.

green algae, is a plausible hypothesis. But the plant drawn by Turpin
is not a simple unicellular alga, whether filamentary or foliaceous. It
is, instead, a glossary plant, an impossible concentration of the most
diverse forms that the single anatomical parts of a plant or flower
can assume. How can a leaf be fashioned? It can be simple or compound. If simple, it may have an entire, serrate, or dentate margin. If
it is compound, it may be digitate or pinnate. If pinnate it can com-
prise an even or uneven number of leaflets. As far as the root is concerned, it may be large and meaty as in a carrot (taproot), or slender
and branching as in wheat (diffuse root). The plant will then be able

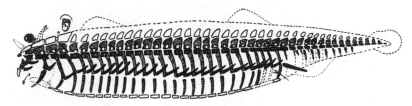

Figure 3. Archetype of a vertebrate according to Owen.

to produce tubers or rhizomes, tendrils, and stolons. Its flowers may emerge from the leaf axils, but could instead occupy the apex of the main stem or of a lateral branch. All this, and much else, finds its place in the dense and multiform architecture of Turpin's *Urpflanze*. I doubt that Goethe (who, at the time this table was printed, had already been dead five years) would have recognized himself in this caricature of his morphology. Maybe he would, however, have appreciated the archetype of the skeleton of vertebrates that Richard Owen published about ten years later (fig. 3).

A Skeleton for Everyone

The title of the work in which we can find this drawing is significant: *On the Archetype and Homologies of the Vertebrate Skeleton.* It is precisely to Owen that we owe the introduction of the concept of homology into biology, in an interpretation that—deprived of its original idealistic connotations, which we shall discuss very shortly, and proposed instead as part of an evolutionary reading of the living world—will continue basically unmodified to this day.

A similar concept, certainly, had also been proposed by other authors, a little before Owens's time, and by none other than Geoffroy Saint-Hilaire. But it is important to observe that, if one wants to avoid misinterpretation in reading the works of these authors, the concept that Owen associates with the word *homology* had, by his predecessors, often been called *analogy* (a term that for Owen, and for us, has a very different meaning). The occasional use of the word *homology* by these authors has a meaning that must be evaluated on a case-by-case basis, often corresponding to what we today call *serial homology* (the correspondence that exists, for instance, in a terrestrial vertebrate between its humerus and its femur, its hand and its foot).

What definition does Owen give for the concept of homology? In the *Lectures on Invertebrate Animals* of 1843 he defined as homologous "the same organ in different animals under every variety of form and function." In some ways, this can be regarded as a good definition. In fact it underscores the irrelevance of the form, as well as the function, of the structures being compared. Our upper limb is homologous to the upper limb of a mole and to the wing of a bat, even if its shape is very different from both of them, and it certainly is not an ideal appendage to use for excavating without the aid of a shovel, much less for flying.

There is however in the definition that Owen proposes, a rather problematic element: what does "the same" mean? On what basis can we affirm that a specific organ of animal A is "the same thing" as a specific organ of animal B? Certainly a researcher in comparative anatomy would never dream of looking for possible homologies between the ear of a mammal and the liver of a bird. And not only would such a wild comparison hardly ever be undertaken, but in the course of a comparison between those anatomical parts that are more obviously comparable to one another, as for instance of a bone with another bone, specific criteria come into play that already had found their place in the mind of zoologists before Owen's time: the criterion of relative position, for example.

Isn't it suggestive that, as we proceed along a bird's wing, starting with its articulation relative to the animal's trunk, we find, in succession, just as in a human being's upper limb, a segment supported by one single bone (humerus), followed by a segment supported by two bones (radius and ulna), and then those bones that, in this order, form the carpus, the metacarpal bone, and the phalanxes of the hand's fingers? It was probably precisely this sort of underlying geometry that made Owen first design a basic model, of which any vertebrate skeleton would then be a variant. An ideal model, an archetype. A model that was, certainly, much more realistic than Turpin's *Urpflanze*, but still the product of a generalization that, as such, does not correspond exactly to any real living being, of yesterday or today.

An Exemplary Crayfish

Much water has flowed beneath the bridges since Goethe, Spix, Turpin, Owen, and Geoffroy presented their comparative exercises, in words or images, to the world. And though traces of a worldview that we perhaps justifiably criticize in their works are still among us, it would be premature and facile to declare them obsolete.

The mollusk type, for example, displayed in most zoology books, is an archetype—it has the features of a snail, or perhaps better a limpet, at least in the shape of the head and the foot, but differs from such by virtue of the linear arrangement of its internal organs, with no trace of the torsion that makes the anatomy of gastropods characteristically complex. The models of bacterial, animal, and plant cells in books on general biology are also archetypes, as are those of flowers (sepals, petals, etc.).

One could object that it would be very difficult to do otherwise. In fact, without such a model, without obvious generalizations, it is almost impossible to introduce the basic notions of morphology, and not only morphology. Allow me to dissent, using as an example a precious tome by Thomas Henry Huxley, whose first edition appeared in 1879. Its title is *The Crayfish*, and though it describes the European freshwater crayfish, the true nature of this work (almost four hundred pages) is better revealed by its subtitle: *An Introduction to the Study of Zoology*.

There is no contradiction between title and subtitle. Huxley devotes most of the book to the crayfish, but his true objective is to introduce the reader to the great problems of biology. The crayfish is, in other words, a means for discussing embryo development and metamorphosis, the physiology of respiration and mechanics of movement, biogeography and phylogeny. It is also a defense of evolutionary theory, an attack on finalism, and even a minor discourse on epistemology. All this by means of the river crayfish? Actually, yes, all this and more for the author explores not only one, but several species of freshwater crayfish, thus giving him the opportunity to include a chapter on biogeography and phylogeny.

In Huxley's book there are no archetypes. The eighty-two illustrations depict only real objects, even with the simplification that drawings naturally introduce. And yet, when the readers have reached the last page, they have become experts not (only) in crayfish, but have also acquired (above all) an organic set of notions of animal biology which they can also apply to animals that differ greatly from the crayfish. In other words, here is a general zoology told by and large by means of a single species.

One can say that medical students follow a similar path, since the different aspects of biology (anatomy, physiology, embryology, etc.) are also presented to them by means of a single species, in this case the human being. So, if my contention (or Huxley's) is legitimate, one could conclude that a medical curriculum prepares one for the profession of zoologist better than a curriculum in which the student is confronted with the archetype of mollusks or the abstract model of an animal cell. It is not difficult, naturally, to distance oneself from such objections. And, to remain on now familiar terrain, I would invite my adversary to read *The Crayfish* from beginning to end. S/he would soon realize that it is one thing to write about one species (be it a human being, another animal, or a plant) because it is only about that one that you and your readers are interested, and quite another thing to use it simply as a familiar and accessible example, a starting point from which to undertake appropriate comparisons and generalizations.

Archetypes and *scala naturae* are not, unfortunately, the only heritage that survives today in a biology that has its roots in pre-Darwinian times. There are also, especially in developmental biology, some forms of finalism that continue to impede the formulation of simple and direct lines of inquiry. Responses to the latter could clear up many obscure points that concern many important aspects of the organizational designs of living organisms and the developmental processes that give them origin. More about these issues later.

Chapter 3

Easy Numbers, Forbidden Numbers

———————— ❖ ————————

The Scolopendra's Legs

Scolopendras are robust terrestrial invertebrates with a poisonous bite, present mainly in the tropics. In the United States there are four species of the genus *Scolopendra*. However, tropical species are more numerous. When it hatches from its egg, the small scolopendra has twenty-one pairs of legs, a number that remains constant even as it undergoes a series of molts similar to those of the caterpillar of a butterfly. This number—twenty-one pairs of legs, attached to as many segments of the body—is common to the various hundreds of species of scolopendras spread around the entire globe. There are, however, exceptions to this rule. A sizable minority of scolopendras is equipped with twenty-three pairs of legs, not one more, not one less. But these are different species. It seems that only in one rare South American species, individuals with twenty-one pairs of legs and those with twenty-three pairs of legs coexist.

That the number of segments (and legs) is invariable within each species is interesting in itself. In fact it suggests that the (genetic?) machine responsible for the production of these segments functions precisely. The most unusual aspect of this situation, however, is the complete absence of scolopendras with twenty-two pairs of legs. No one has ever seen such a specimen, not even an anomalous individual within a species that normally possesses either twenty-one or twenty-three pairs of legs. In truth we are discussing animals that have not aroused much interest and have not been extensively studied, and so it is possible that some scolopendra with twenty-two pairs of legs, yet to be discovered, may exist in some remote area of our planet. However there are excellent reasons to believe that the existence of such an animal is highly unlikely. Good evidence to support this claim comes from geophilomorphs, a group of small animals similar to scolopendras in which the trunk, very elongated and sometimes even wormlike, is provided with a greater number of legs, from a minimum of 27 to an ascertained maximum of 191 (fig. 4).

To begin with, please note that these two numbers (27 and 191) are also uneven, as are twenty-one and twenty-three. But we should immediately add that even numbers are also impossible in geophilomorph centipedes. In geophilomorphs, however, there is an even more paradoxical situation. On the one hand certain species exhibit a perfect uniformity in the number of segments, even if these are, for instance, forty-five or forty-nine. On the other hand, the number of trunk segments in this group is variable in most species, but here as well, even numbers are impossible.

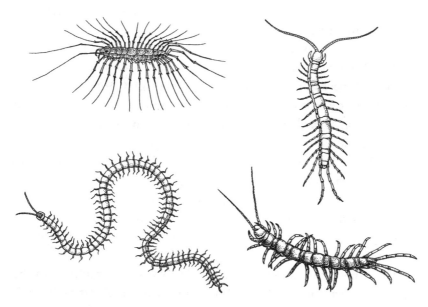

Figure 4. Centipedes.

The fact that an animal produces exactly forty-nine segments, without individual exceptions, makes the hypothesis that these segments are produced serially, one after the other, unlikely. What segment-counting device could the geophilomorph possess, to always stop the production of segments precisely at the same point? Perhaps, one might think, some abnormal individuals do develop, but natural selection is very efficient in eliminating individuals that exhibit one segment more or less than is normal. Consequently all the circumstances that can make the mechanism for producing segments more precise are favored, and the probability of running across a deviant individual is so low, that, in fact, none have ever been seen.

It is more likely, however, that the situation is different. The "explanation" in terms of natural selection for the apparent absence of deviant individuals as regards number of segments in a species in which the rule followed by all individuals is to produce exactly forty-nine, through seeming very much ad hoc, is still marginally plausible. This margin, however, decreases to zero once we attempt a similar explanation to account for the total absence of individuals with an even number of segments in those species of geophilomorphs in which the number of segments is variable even within the same species, or rather, within each single population.

How can we explain, for example, the total absence of individuals with forty segments, or with forty-two, in a population in which some individuals have thirty-nine segments, others forty-one, and still others forty-three? Is it possible that geophilomorphs with forty pairs of legs function so badly (for instance, that they have such serious problems with movement) that they are literally erased from the face of the earth? More than improbable, this hypothesis seems laughable. Who can believe that a geophilomorph with forty pairs of legs will be condemned to trip up ruinously among the clods of earth, while his brethren with two more or two less legs, happily frolic around, transmitting the genes for their victorious number of segments to numerous offspring?

This whole story about even and odd millipedes might seem the futile argument on which a zoologist incapable of elevating himself to the level of the great problems of modern biology wastes his time, but this is not the case, for the story of these geophilomorphs has wide-ranging implications. It invites us to reflect carefully on the role of natural selection as a principle on whose basis we attempt to explain the diversity of living things. A hypothetical example will help us formulate our questions more clearly. In this case to avoid the abstract and dry language of modern biology, I will ask the reader's forgiveness for "personifying," if only for a couple of paragraphs, nature and natural selection, in their doing and undoing, and in pointing to the ties that bind them, just as they bind human beings. Please do not take these formulations literally.

It may be, some will say, that nature has never been able to produce butterflies with six or even eight wings. In other words it is likely that natural selection never had to busy itself with such occurrences, even though it is probable that it would have found these forms less efficient than the usual four wings that our butterflies use to fly. One must say, however, that for fast flight even four wings might be too many. The possible errors are reduced if the two pairs of wings can beat in a perfectly synchronized manner, and in fact many butterflies and moths have devices to solidly hook together the forewing and hind wing on the same side. It is as if the small flying machine has, to all intents and purposes, a single membrane on each side. In other insects such as Diptera, the order to which mosquitoes, fruit flies, and bluebottles belong, the functional wings, which the animal depends on for its movement through the air, are in fact reduced to two, and the rare individuals with four wings that have occasionally been found in laboratory fruit-fly hatcheries should be considered occasional mutants that natural selection inexorably eliminates at their first appearance. Returning to butterflies, however, it is probable that the scope of natural selection has influenced the size of the wings, their precise shape, the color of the scales that cover them, and the thousands of ways in which these colors are combined in the shape of eyes, band, or festoon. But butterflies with six wings, or with eight, have remained in the limbo of unrealized forms, whether they are possible or not.

The Nonexistent Variant

But what can one say about a geophilomorph with forty pairs of legs? Is nature perhaps incapable of producing one, inside a cozy family in which the other offspring exhibit thirty-nine or forty-one segments? Here is precisely where the issue lies. If nature were to produce geophilomorphs with an arbitrary number of segments, the total absence of individuals with an even number of segments could be attributed to natural selection, but it would have to be extremely efficient. It is therefore more probable that the curious discontinuous distribution of the "possible" number of segments, within the domain of the geophilomorphs, has some other cause. In other words it is probable that individuals with an even number of segments don't exist because it is not possible to construct them. If this is how things stand, natural selection does not eliminate them; it simply never has to confront them, just as it does not have to confront butterflies with six or eight wings. This in turn means that the explanation for the fact that geophilomorphs, or more generally centipedes, always have an odd number of pairs of legs must not be sought in the greater ability to survive (i.e., ultimately in their greater reproductive success) that individuals with an odd number of pairs of legs possess when compared with those that have an even number, but rather in the fact that nature "knows how" to produce the former but apparently not the latter. A centipede with an even number of legs is more than a monster: it is an impossible monster, almost.

An interesting conclusion can be derived from this dual story of butterflies and centipedes. When we hypothesize the existence of two similar animal forms, only one of them actually existing in nature, the absence of the other is not necessarily to be attributed to natural selection. In other words the explanation is not necessarily to be sought in the better adaptation of one form over the other to the conditions of the environment. Instead it is possible that our attention should primarily shift toward those mechanisms of development that are responsible for the generation of living organisms. These mechanisms are capable of producing, with minimal variations in the initial conditions (the inherited genetic information, but only this?—we shall soon return to this topic), both those forms that will be successful, and those that will be erased by natural selection. But these mechanisms are not omnipotent. On the contrary sometimes they don't seem to be able to modify their current "mold" even slightly. They are not capable, for example, of adding a segment to a centipede. When it seems as if they were able to, they always "err" on the side of excess, adding two instead of one. But this description only applies if it is a question of adding segments, rather than something altogether different. That the latter possibility may be the case is suggested by the study of other little creatures: leeches.

The Leech's Segments

Leeches belong to the zoological phylum of the annelids, together with earthworms and polychaetes, a numerous and heterogeneous group of wormlike animals that almost all live in marine environments, and are equipped with numerous pairs of short nonarticulated appendages at the sides of their body. The name these animals have been assigned expresses their segmented appearance well: along the body's main axis there occur a series of, generally uniform, repetitive units that have come to be designated as segments. In the most common species of our regions' earthworms, the body is formed by about one hundred segments, usually not many more. In polychaetes this number can vary from only a few units to more than one thousand. In leeches on the other hand we are confronted with a rather peculiar situation: the number of segments is rather low and absolutely constant, and it never changes once the animal has hatched from the egg. In other words there are no differences either within an individual species (for instance in the medicinal leech, which was once sold in pharmacies and was used to let blood), or between any of the more than five hundred known species. One should add, for the sake of exactness, that it is useless to look for differences between the sexes since, as in the case of earthworms, leeches are hermaphroditic.

How many segments, then, make up the body of a leech? A superficial observation could lead us astray, or into uncertainty. The rear end of its body in fact exhibits a sucker, whose segmental composition is hard to determine. In addition, the body's surface seems to be subdivided into a large number of rings: at least sixty, but often a much larger number. A dissection, however, is sufficient to help us answer our question. At most, the remaining uncertainty will be on the order of one or two segments. A comparison with an earthworm or another annelid with a simpler structure (more specifically, one without a posterior sucker) can help guide us. In an earthworm in fact it is simple to verify that in each body segment there is a group of organs that is repeated, basically without any variation, in all the other segments. Of this group of organs, those that preserve their serial distribution even in annelids with a more specialized structure, like leeches, are the ganglia of the central nervous system: one pair per segment (one on the right, one on the left) in a ventral position, in other words beneath the intestine. In the posterior region of a leech's body, that is, in the area of the sucker, the ganglia are so close to one another as to basically seem fused together, but their original number can still be discerned. On the basis of the number of ganglia and a few other internal structures, it is therefore possible to establish that the total number of segments that make up the body of a leech (of any leech) is always thirty-two. In older zoology treatises this number was often also indicated as thirty-three or thirty-four, but there seems to be no doubt on the need to rectify the older estimates.

Unexpected Arithmetic

Once again, these numerical issues might seem to be subtleties of relevance only to a rather pedantic specialist, but this is not the case. Thirty-two is in fact a very different number from thirty-three— thirty-two is also two to the fifth power, while thirty-three only has three and eleven as divisors. Having established that thirty-two is the number of segments that constitute the body of a leech might in fact suggest an extremely simple mechanism, which, if actually utilized during the embryonic development of these annelids, could easily account for the precision and invariance with which this numerical value is reached. It is in fact easy to imagine a mechanism based on a repetitive series of binary divisions that would initially divide the embryo into two parts (anterior and posterior), each of which would in its turn then be divided into two, by means of a similar process, thus bringing the total number of segments to four. Three subsequent binary divisions, applied in a uniform manner to all the segments obtained from the preceding divisions, would take us, in that order, to stages with eight, sixteen and—finally—thirty-two segments. A simple and trustworthy mechanism, which could therefore account for the constant number of segments to be found in leeches. A pity therefore that these animals actually form their segments in a quite different manner. And it really does seem that they are able to "count" the segments they produce, one after the other.

The production of segments in leeches is tied to the activity of specific cells called teloblasts. They behave rather like stem cells, or better, like "finite-life stem cells." In a very precocious phase of development, when one is barely able to distinguish the body's main axis, ten specific cells, called teloblasts, start to proliferate: five on the left side of the animal and five on the right. Teloblasts are rather large cells that are subject to a repeated series of asymmetrical divisions. The two products of each mitosis, in other words, have a different destiny: one of the two cells preserves the characteristics of a teloblast, and will therefore undergo another division of the same type, while the other will become the progenitor of a homogeneous group of cells, at least initially. They will occupy a specific position inside the embryo and will also have a very specific destiny.

In the case of the leech's ten teloblasts, a first division will give rise to an equal number of cells that will then form the body's first segment, in addition to an equal number of teloblasts that are ready to divide a second time, giving rise to a second series of cells, progenitors of the cell population of the second segment, and so on.

All these teloblasts undergo division more or less synchronously with one another. Division after division, each time the teloblasts leave a group of cells behind that is destined to form a new segment, they move progressively toward what will become the body's rear end. The reason the teloblasts' activity ceases after they have generated a number of offspring cells sufficient for the formation of thirty-two segments is as yet unknown. It is known, however, that the proliferation of each teloblast is an event that is basically autonomous from the proliferation of the other teloblasts. And it is not even as precise and rigorous a process as one might expect, judging from the final result. In fact the number of times each teloblast divides is not exactly thirty-one (as many times as would be necessary to arrive at the fateful number of thirty-two segments), but it is usually higher and rather variable. One arrives at thirty-two segments, neither more nor less, only at a later stage, with the elimination of the youngest cells produced by the teloblasts, thanks to mechanisms that are however not yet known.

To sum up: the formation of segments in leeches is an extremely precise process, at least in its results, even though it does not occur thanks to what might seem to us the only safe and easy way to arrive at a final number of units that is a precise multiple of two. On the other hand this number (thirty-two) seems impervious to the process of natural selection. There are leeches with a very elongated and cylindrical body resembling earthworms or little snakes, and leeches with a short and flattened body resembling a leaf. In both cases, the number of segments that constitute the body is always the same, that curious number thirty-two, which these annelids replicate in their development—and there seems to be no exception to the rule. This number seems to express limitations arising from developmental mechanisms, rather than a choice due to natural selection.

And yet it is not quite true that a certain kind of flexibility is totally absent from leeches, at least in terms of repetitive units. A little earlier we mentioned how the body surface of these animals is articulated into a variable number of "rings." These are obvious but superficial subdivisions that only affect the body wall and have no equivalent in the internal anatomy of the animal. There is however a very precise relationship between the number of segments the body is comprised of (the famous thirty-two fundamental units with their corresponding pairs of ventral ganglia) and the number of external rings. This correspondence is definite and constant for the segments belonging to the middle section of the body in all species of leech; but it is not the case for the anterior segments, where the number of rings per segment is progressively reduced, until it is only one for each of the terminal segments, and finally, on the posterior sucker, there is no trace of any rings at all.

Let us therefore devote our attention to the intermediate segments. In many species of leeches, each of these segments is articulated into three rings; in many others, including the medicinal leech, into five rings. These are the most frequent forms, but in some genera the number of rings per segment climbs to six, seven, or even fourteen. And, as we could perhaps rationally expect, the number of rings per segment is lower in species with a short leaflike body and higher in those with a long, cylindrical body.

One could say that nature has found a way to free itself from the bond of a very rigorous developmental mechanism that generates series of segments according to a fixed number. If there is no variability in the number of segments, why not subdivide each segment into a variable number of lesser units? Naturally we should expect that the number of rings produced per segment should also be subject to limitations. In fact as we just saw the majority of leeches have either three or five rings per segment. But in leeches, even numbers of rings, such as two, four, and six, although rare in nature, can still occur. In this case at least, it is legitimate to ask if the preference leeches seem to exhibit for an uneven number of rings per segment depends exclusively on the greater ease with which these subdivisions can be realized (in terms of developmental mechanisms), or if there is, perhaps subordinately, some adaptive advantage, maybe connected to locomotion. This "renewal" of variability by means of a mechanism—the subdivision of each segment into several rings—that adds a new dimension to a structure that would otherwise become fairly rigid because of an all too rigorous developmental mechanism, is a situation that we shall come across again in these pages. For the time being, however, we shall do well to examine another situation in which an apparently extremely rigid rule actually allows for exceptions, even if rare.

The Giraffe's Neck

Since the times of Lamarck's *Philosophie Zoologique* (1809), the giraffe seems to have found its legitimate niche in works that discuss biological evolution. For the old French naturalist it offered an excellent example of the effect of a prolonged, continuous use of an organ. In this view, an extremely long neck would have been the result of the efforts of generations of giraffes attempting to reach the leafy fronds of acacias, a precious source of food once the dry season has withered all the savanna's grasses. The corresponding Darwinian interpretation, in terms of natural selection, hypothesizes inherited individual differences in the length of the neck, within past and current giraffe populations, with a corresponding advantage for individuals with a longer neck. The end result would therefore be caused by imperceptible increments due to the fact that individuals who already possess the advantage of the longer necks will be more likely to reproduce, transmitting this characteristic to their offspring.

To say that, in the light of modern biology, the Darwinian scenario is credible whereas the Lamarckian is not, does not exhaust the problem of the origin of the giraffe's neck, nor does it exhaust our interest in it. In fact we still have to ask ourselves about the nature of this intraspecific variability on which natural selection has been able to operate over the course of generations.

It would be easy to suppose, for instance, that the length of the neck is somehow tied to the number of cervical vertebrae that constitute its skeletal scaffolding; but this is not the case. In fact we only find seven cervical vertebrae in the giraffe's extremely long neck, the same number we find in the neck of human beings, and, more generally, in almost all mammals. In other words between one species and the next we find only differences in the form, not the number, of the cervical vertebrae. The giraffe's are much more elongated, ours are much shorter—that is all. Therefore in terms of skeletal anatomy the evolution of the giraffe's neck has only entailed a progressive elongation of the seven cervical vertebrae, because only this aspect (the shape of the individual vertebra) is subject to variation. Natural selection has no influence on the number of vertebrae in the neck of mammals, or, better, it has no materials on which to exercise such influence.

Indirect Effects

To understand something about the invariance of the number of cervical vertebrae in mammals we must, once again, refer to developmental biology. We would do well, however, to examine the entire spinal column instead of restricting our attention exclusively to the cervical vertebrae, which represent its anterior section, that closest to the head. In addition we need to learn something about the manner in which genes express themselves in the course of development. We shall therefore embark on a very brief digression on this topic.

All the cells that constitute an animal's body are endowed with a nucleus that contains a copy of the entire inherited genetic information (or genome) of the individual, inscribed in the specific structure of the long DNA molecules. According to traditional descriptions (which modern molecular biology views as rather crude and imprecise, but here they should suffice), the genome is articulated into units, the genes, each of which can be expressed separately from all the others, within the cell. The expression of a gene takes place in two principal phases, transcription and translation. Transcription consists of copying the segment of DNA corresponding to the gene by constructing an equivalent segment of messenger RNA (mRNA) that will move to a place appropriate for translation; in other words for the construction of a specific protein, which in its structure mirrors that of the specific mRNA (and, therefore, of that specific gene) from which it takes its information.

Even though all an organism's cells possess a complete copy of the entire genome, only a part of it is expressed in each cell. The differences that exist, for example, between a muscle fiber and a liver cell, are precisely a function of the different choice of genes that are expressed in each cell.

A gene's expression is diversified not only in space, that is, between one group of cells and another, within an embryo or an adult animal, but also in time. In fact some genes are always expressed, others are expressed in a lasting fashion only starting from a certain point in development. Especially interesting are some groups of genes that are only expressed during a more or less precocious, specific phase of embryonic development. These genes may be of great importance in the realization of the fundamental traits of the body architecture of an animal: for example in specifying which shall be the back and which shall be the belly, as in the case, previously mentioned, of *chordin* and *Bone Morphogenetic Protein-4*.

Among the many genes involved in these fundamental phases of embryonic development, a special place should be reserved for the so-called *Hox* genes, to which we shall devote particular attention in the next chapter. It should be sufficient to know, to be able to immediately return to the problem of cervical vertebrae, that each of these *Hox* genes expresses itself in a more or less limited area of the animal's main body axis and that the anterior limit of its domain of expression represents a sort of "molecular marking" that specifies a precise position within the embryo. Each of these positions specified by the localized expressions of one or more *Hox* genes can then be "interpreted" by other genes whose expression will then lead to the realization, at that precise location (or starting from that location and then proceeding backward), of a particular organ.

Returning to the mammal's vertebrae, or those of any other vertebrate, these are formed in a serial manner, one after the other, beginning with the first cervical vertebra. The passage from one section to the next (from the cervical to the thoracic, from the thoracic to the lumbar, from the lumbar to the sacral, and, if necessary, from these to the coccygeal/caudal) is not marked by interruptions and renewed starts in the production of vertebrae, but rather by the periodic appearance, at the moment in which the different vertebrae are produced, of some proteins that are encoded by as many genes belonging to the *Hox* complex. It is precisely the presence of these molecules that directs the differentiation of vertebrae toward their destiny, becoming, respectively, cervical, thoracic (etc.) vertebrae.

It is therefore entirely plausible that the number of vertebrae belonging to each section, and to the cervical one in particular, is fixed, more or less rigorously, not by virtue of a specific mechanism preserved by natural selection because of its precision in producing a specific number of vertebrae, but only as a secondary consequence of the appearance of the products of the various *Hox* genes during moments of embryonic development that are rigorously preserved for an entirely different reason. What this reason might be is difficult to tell, since the sequence of events during embryonic development, and the genes that are given expression at this time, generally give rise to a cascade of effects across a wide spectrum. One can, however, indirectly verify the plausibility of this hypothesis and, therefore, of the opportunity to invest significant research efforts in this direction. If it is true that the transition from the production of cervical vertebrae to the production of thoracic vertebrae takes place during a delicate moment (and therefore, given its temporal position, one not easily modifiable) of embryonic development, we should expect that any anomalies in the spinal column localized in precisely that position (for instance, an exceptional case of eight cervical vertebrae in place of the usual seven) might be associated with other disorders, perhaps even important ones. And this is precisely what Frietson Galis and her collaborators have found in an accurate review of the possible developmental anomalies that can affect the human species. Those anomalies associated with "errors" in the development of the normal series of seven cervical vertebrae are generally very serious, though not as a direct consequence of this skeletal anomaly, but rather because of pathologies of a very different nature that are associated with it. These are, most likely, the target of a selection that, among its indirect effects, also fixes the number of cervical vertebrae in all mammals at seven.

Or, to be more precise: of *almost* all mammals. The exception is represented by two families, comprising a handful of species. Of the four living species of Sirenia (large aquatic animals related to elephants), the three species of manatees have only six cervical vertebrae, while the fourth species, the dugong, usually has seven, though individuals with eight have also been found. Analogously, among the edentates, the two species of two-toed sloths have six cervical vertebrae, while the three species of sloths with three toes have either eight or nine. In both groups, therefore, one finds a subversion of the rule that otherwise applies to mammals: here, within the same family, there are species with a shorter and others with a longer series of cervical vertebrae than is normal. Their number can even vary within the same species. How and why nature has allowed itself these exceptions in these two groups would be interesting to know. For the moment, however, no specific study on the topic has been undertaken.

On the Fingers of One Hand

Whatever the mechanism responsible for their formation may be, the segments of a scolopendra's or a leech's body and the vertebrae of a vertebrate have various characteristics in common. Disregarding the fact that in all three cases we are dealing with repetitive units that articulate the trunk of the animal so as to aid in its locomotion, the series also share the fact of comprising a more or less large number of similar elements, at least within a specific region of the body. In the case of the scolopendra the last pair of legs is significantly different from those that precede it, but the other twenty (or twenty-two) pairs are almost identical. In the leech, as we saw previously, most segments exhibit the same structure and division into rings, the only exceptions being those segments closest to the body's extremities. And in the case of vertebrates, the vertebrae of each region (the cervical, the thoracic, the lumbar, the sacral, and the coccygeal/caudal) are similar to one another. Other series of repetitive structures, however, show a greater degree of complexity, in the sense that their constitutive elements all differ from one another, as happens for instance in the fingers of terrestrial vertebrates.

Our hand, for example, has five fingers whose structure is fairly uniform, with the exception of the thumb, which has only two phalanges—all the other fingers have three. Each finger, however, has its own identity, which is not due only to its position (from the first to the fifth finger, proceeding from inner to outer), but also to morphological differences that are more obvious in the case of the thumb and the little finger, and also allow us to distinguish index, middle, and ring fingers. The same situation, roughly, is repeated in the case of the foot, except that our two-handed life makes us pay less attention to the morphology of the lower limbs. Five fingers can be found in many other mammals; five can be found in many reptiles; the hind limbs of almost all species of frogs and salamanders also exhibit five toes. Birds and numerous other terrestrial vertebrates have four fingers/claws; still others have only two or three, until we reach a leg with a single toe, the hoof, which is typical of donkeys and horses. No living vertebrate has more than five fingers per limb, but a slightly higher number (seven or eight) was exhibited by some very ancient forms (*Acantosthega* and *Ichthyostega*), that were undoubtedly "close" (in geological terms) to the first forms of vertebrates that started to rest the weight of their bodies on four limbs.

No one has ever doubted that the five toes of the hind limb of a frog correspond, one for one, with the five toes of our foot, and it is quite possible that such a correspondence exists, but we cannot treat this hypothesis as if it were a certainty. The reasons for looking deeper are due to the surprises encountered in the course of other comparisons. In fact we can ask ourselves which toes are present in those legs with less than five toes. For example, which toe constitutes the horse's hoof? Which finger/claw went missing in birds?

Behind these apparently easy and innocuous questions, some issues of considerable importance for the entire field of comparative biology lie hidden. Let us start with our hand and its five fingers, each one different from the next. How can we imagine a series of hands with a progressively decreasing number of fingers? Probably the first solution that comes to mind consists in imagining that four of our fingers gradually disappear: perhaps the little finger first, followed by the thumb, so as to balance the first loss, while at the same time preserving a continuous series, even if it only consists of three fingers. Then it will be the fourth finger's turn, followed by the second, leaving the middle finger in its position, as the last survivor. This is actually the order that is most often invoked to account for the progressive series of reductions in the number of fingers/toes, as they have been observed in some groups of mammals and lizards.

A reconstruction of this kind presupposes that each finger preserves its identity even when the limb has less than five fingers, but this may not always be the case. In other words it is not necessarily true that the two fingers on a limb that only has two are equivalent to the two fingers on a limb that has five: i.e., that it is possible to assign to them the value of second and third finger. Ultimately, the positions (from first to fifth) are not absolute, but relative, and it is possible that the mechanisms that give each finger a precise identity do not depend only on the different expressions of genes that, on a case-by-case basis, may be expressed or not (where there is a thumb or not, for instance), but also on information about position, which, to put it simply, may not find a correspondence between a situation with five fingers and one with two. To put it differently: in order to have a thumb, an index, middle, ring, and little finger, it is not sufficient to have a set of genes expressing themselves in a differentiated fashion from one finger to the next, endowing each finger with its precise identity. One also, and perhaps above all, needs to have fingers in the precise sequence of positions that allows these genes to express themselves differently from finger to finger. If a hand only has two fingers, and this is not due to the accidental loss of the other three, but because all the material available during embryonic development for the expression of fingers is utilized to make two fingers instead of five, there is no more certainty associated with the relative positions. In the abstract, two fingers will not correspond, in terms of position, to any two specific fingers of a hand with five fingers. In this case the only possibility for a rigorous comparison is between the set of five fingers from the one hand with the set of two from the other. The identity of the single finger is put in question because of the overall reduction of the limb.

The actual significance of these reflections, very important from a theoretical point of view, has to be established on a case-by-case basis. Conditions in ruminants, for example, whose limbs are at most endowed with four functional fingers/toes, and, more often, with only two, do not seem to be controversial. In the first case it seems as if the first finger is missing; in the second only the third and fourth fingers survive. Other situations, however, are more controversial and prove that it is necessary to undertake these comparisons with great caution.

Number and Position

This is the case with birds, for instance. As previously mentioned, their feet only have four fingers/claws. And in their wing one can only see three. Is it possible to assign each of these a specific identity? In recent years this problem has been the subject of a heated debate, because of its importance in determining the kinship between birds and dinosaurs.

According to many experts, birds are basically a specialized branch of dinosaurs, whereas others believe the degree of kinship between the two groups is much more remote. In any case it seems legitimate to believe that dinosaurs' forelimbs ended with the first three fingers/toes, but this does not necessarily mean the same is also true for the birds' wings. In fact there are those who believe that these three fingers correspond instead to the second, third, and fourth unit of a conventional limb with five fingers. In fact if on the one hand the appearance of these three fingers seems to support the interpretation that sees them as fingers one to three, on the other hand their position relative to the carpus seems to support their being designated as fingers two to four. Because of the reasons given above there is still some margin of doubt as to the legitimacy of such a precise identification.

In this case let us admit that a comparison at the level of an individual finger is justified. The fact remains that the criterion of position would lead us to different conclusions compared to the criterion of the specific identity of each finger. Perhaps, however, there is a way out. According to Günter Wagner and Jacques Gauthier, the opposing conclusions to which the criteria of relative position and that of the special quality of each finger respectively lead us, don't necessarily contradict each other. In fact the developmental processes that lead to the construction of the fingers are distinct from those that endow each finger with its characteristic identity. It is therefore possible that the three fingers in the wings of birds really do represent fingers two to four as far as the process of *the fingers' formation* is concerned. Each finger however, would then see an "interpretation" superimposed on it, that would concern its *specific quality* and assign the value of first finger (the equivalent of our thumb) to the first available finger, even if in fact, in terms of original position, it is the second finger. And the same would be the case for the remaining two fingers.

In the next several years we shall see if, perhaps, this complex interpretation is closer to the truth. But even if, in this specific case, it should turn out to be mistaken, we would still do well to remember the more far-reaching suggestion it entails. When a brief series of parts is constituted of elements that are similar but yet differentiated from each other, as in the case of the five fingers of our hand, it is easy to forget that the developmental mechanisms responsible for the formation of these parts, in the numbers we observe, are generally distinct from the mechanisms that endow each of these parts with its characteristic identity. It is one thing to construct five fingers, another to construct five different types of finger. So long as the numbers coincide (the number of parts and the number of types of parts), the existence of the two distinct mechanisms can remain hidden, and this can become a significant obstacle to the understanding not only of the developmental process itself, but also and above all of the paths along which these mechanisms, and their products, can evolve.

At the same time, one should expect a sort of co-evolution of the mechanisms that determine the (maximum) number of parts produced and the mechanisms responsible for the number of different types in which they can be differentiated. In fact, as we mentioned earlier, the first terrestrial vertebrates were endowed with a number of toes that was slightly higher than the five that soon became the highest number that could be reached by the toes of a normal leg. But *Acantosthega*'s or *Ichthyostega*'s seven or eight fingers/toes were not differentiated into as many different types: one certainly could not find more types than in the leg of a modern vertebrate. No vertebrate has learned how to construct more than five types of fingers, and the mechanisms that produced them soon conformed to this number, "renouncing" the construction of a greater number.

Once the architecture of the pentadactylic limb had been consolidated, some other possibilities still remained. The most obvious one, and the most widespread, consists in the reduction of the number of fingers, as mentioned above. Another possibility, much less widespread, but not less interesting, consists in the addition of a sort of "false accessory finger" to the real fingers, whose number does not seem to be able to exceed five, at least in modern vertebrates. This is the case in what appears to be the first finger in the hand of the giant panda, a finger whose bone scaffolding is constituted by a sesamoid bone that is articulated onto the metacarpal bone aligned with the first finger. The situation is not very different for the mole, where a noticeable sesamoid, the falciform bone, is in an analogous position in front of the thumb. Six fingers therefore; or better, five plus one.

Chapter 4

Privileged Genes

———————◆◆◆·———————

Unity of Body Plan

Since Belon's times, one of the fundamental criteria that has always guided zoologists when comparing animal forms has been that of relative position, which supports us when, for instance, we say the wing of a bird is equivalent to a human being's upper limb. Both appendages in fact, present easily comparable points of articulation relative to the spinal column. Within each articulation, as we already mentioned, we find the same succession of bones, starting with the humerus, which is followed by the radius/ulna pair and, then, by the small bones of the carpus and the metacarpus (the bones of the palm of our hand), to finally end with the bones of the fingers.

One could say that we are in any case still in the realm of the vertebrates. But what will happen if we attempt to compare two much less similar animals, so different in fact that Cuvier assigned them to two different *embranchements*? Will we perhaps still find some correspondences between organs or body parts, at least from the point of view of their relative position?

The answer to this question is largely affirmative, even though one still finds some problematic situations. There is no doubt, however, that many animals, even though they can be assigned to different groups, share a sort of "syntax of the body," in which it is possible to discern a main axis, with the mouth (and possibly the brain) at one of the two extremities. What one will then find at the opposite extremity is a less obvious question, to which we shall return. For the moment let us content ourselves with the determination of a main axis, along which different anatomical parts succeed one another in positions that are in any case comparable, since this axis at the very least (there are a certain number of exceptions, scattered here and there within the animal kingdom) presents us with an easily recognizable extremity, that, precisely, where the mouth is to be found.

From this point of view, a snake and an earthworm do not differ much one from the other, but the presence of a principal axis with the mouth at one end also allows us to compare either one with a crayfish or a snail. Let us leave aside the problems that a starfish, a sea-urchin, or a sponge raises.

Let us however attempt to deepen our comparisons. It is not clear in fact which premises we appear to take for granted and what implications we might deduce from this state of affairs. When we say that the position X, along the body of an animal A is equivalent to the position X' in an animal B, is our statement really based on the fact that in the position X = X' we find that same anatomical part, the mouth for instance? Or, on the contrary, do we say that two anatomical parts (for example a certain fin on fish A and a certain fin on fish B) are equivalent precisely because they occupy that same position, regardless of the organs that we find there?

The question may not seem significant if, for no other reason, because it appears fairly easy to establish other criteria (beyond that of position) to judge whether two parts are equivalent; additionally, it seems absurd to talk about "equivalence of position" in the abstract, without referring to any concrete structure that may occupy a certain position in one animal or the other.

But things are not quite that simple. The anatomical specificity of an organ and its relative position inside the overarching syntax of the body are two characteristics that do not necessarily change together. In some cases, on the contrary, we have good reason to describe two positions as equivalent even though they are occupied, in different animals, by clearly different organs; in other cases, we have good reason to describe two positions as different, in different animals, even though they are occupied by what is for us the same organ. Let us look at a couple of examples.

Among the numerous mutants that have appeared in laboratory *Drosophila* (the fruit flies that played such a big role in the development of genetics at the beginning of the twentieth century and have now found a moment of renewed glory as a choice subject of study in developmental genetics), two types of really "monstrous" fruit fly have been given special attention. These are not disturbing in part because we are not dealing with human beings, or, in any case, mammals; and, moreover, these are not mutants with two heads or with a gigantic cyclopic eye on their forehead. Fruit flies with greater "horror film" characteristics will be discussed a little further on: nightmarish creatures with a bit of eye on a leg, "constructed" in the mid-1990s by a group of researchers lead by Walter Gehring at the Biozentrum in Basel. For the time being let's be happy with fruit flies with four wings or with a pair of legs instead of antennae.

Drosophila is an insect, and among insects, four wings are the rule rather than the exception. Damsel flies and butterflies, for instance, have four wings. *Drosophila*, however, belongs to that vast order of insects (the Diptera) that fly only with their forewings, while the hind wings are modified to a pair of halteres, short appendages, vaguely resembling the stamens of a flower since they are formed by a filament that at one extremity expands to form a bulbous mass. These halteres, which the insect keeps in constant vibration while in flight, and which seem to decisively contribute to the stability of the small animal while it dashes rapidly through the air, in effect occupy the same position that in other insects is occupied by the hind wings: therefore there is every reason to consider them modified wings. Same position, but different development.

Usually at least. Yes, because, when compared to a wild-type *Drosophila*, the mutants do not exhibit excess appendages, or an extra body segment. Simply put, the segment of the thorax that in normal *Drosophilas* bears the halteres is more developed in these mutants (more precisely it is an almost exact replica of the segment preceding it) and, above all, those that should have been the two halteres appear instead as two wings. Two *Drosophila* wings, of course; in fact, a very faithful copy of the wings that are born by the preceding segment. It is as if the thorax of these fruit flies, instead of being constituted of three segments (the first, second, and third), one different from the next, included three segments traceable to only two types, which could be described as first, second, and again second.

One should remember that the second pair of wings that these mutants exhibit is *not* the original pair of wings, in other words that born by the third segment of the thorax of its last four-winged ancestor in the order of the Diptera. It is not in other words the pair of wings that was transformed into a pair of halteres when this group of insects originated.

To describe these supernumerary wings of the *Drosophila* mutant in a sensible manner, it is necessary to distinguish between two factors: position and special qualities of the structure. From the point of view of position, these wings occupy the place usually taken by a pair of halteres. Moreover, since the halteres are modified hind wings, we can also say that the mutant's supernumerary wings are equivalent, *in terms of position*, to a pair of hind wings. An equivalent butterfly mutant would have two pairs of wings, identical to the fore wings typical of the species. An equivalent mutant from the order Coleoptera would have two pairs of elytrons (since these rigid appendages, no longer utilizable for flight, represent the *first*, modified pair of wings that is characteristic of this order of insects).

Another example of substitution of pieces, but always within the bounds of precise laws of equivalence between the parts, is represented by that *Drosophila* mutant in which the two antennae appear to be replaced by a pair of legs—once more, *Drosophila* legs, symmetrical, like the two antennae they are replacing.

If we recall the fact that the two halteres are modified wings (and that a mutation is known which is capable of replacing the halteres with a pair of wings), we are tempted to ask if the substitution of different types of appendages for one another is, in the mutant, overlapping with a more ancient equivalence between two structural types in this case also—if, in other words, legs are no more than modified antennae, or, perhaps, vice versa (fig. 5).

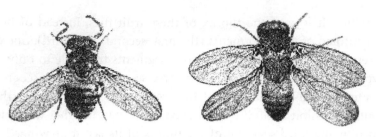

Figure 5. *Drosophila* mutants.

The Universal Appendage

In effect, the idea that one may recognize one unique ancestral model of appendage in arthropods, from which all others would have derived, is dear to many students of developmental biology, and three different versions of this hypothetical model have been proposed. For some the leg represents the basic model; accordingly in addition to the antennae, it would have been the source of mandibles and maxillae, as well as of the male appendages whose function is to transfer spermatozoa during mating, and of the female appendages that many arthropods utilize when laying eggs. According to an alternate model, the antenna would instead represent the primitive appendage. This hypothesis is supported by the fact that the appendages that should differentiate from one another according to their position—turning into jaw, mandible, or leg—all tend to resemble antennae, if, because of a mutation, the expression of one or two genes, whose products are necessary for the appendage to develop as expected, is completely absent in the embryo. There is finally an intermediate hypothesis, according to which the primitive appendage of arthropods was not, in reality, either a leg or an antenna, but something intermediate; and from this intermediate appendage, all the other different types we find in modern arthropods would have derived. It is probable, however, that all of these hypotheses are mistaken. They all in fact start from a presupposition that, from what we know of the developmental history of animals, is almost undoubtedly false.

For an animal to have a series of appendages of different types along the length of its body, at least two conditions must obtain. On the one hand, the animal has to be able to produce several pairs of appendages, which is certainly a widespread but not universal capability. Earthworms, for instance, do not have similar appendages, and neither do ascarids and tapeworms. On the other hand the various pairs of appendages will differ from one another if there are a series of nonequivalent positions, of specialized "local environments," that block the appendages from developing identically along the animal's main body axis.

As far as the origin and evolution of these two conditions is concerned, it is possible to imagine different scenarios. We can suppose for example that the capability of forming pairs of regularly spaced appendages along the body preceded the marking of nonequivalent positions along the main body axis of the animal. Within a similar scenario, it is legitimate to suppose that, at an initial stage, all the appendages produced by the animal were similar to one another, and that only with the progressive appearance of local or regional differences along the trunk, did the specialization of the different pairs of appendages develop. To simplify matters, we can refer to animals whose bodies are divided into segments, like earthworms or leeches, or millipedes, insects, crayfish, and scorpions. We will therefore imagine an ancestor whose body is formed by identical segments, devoid of appendages, from which forms of life whose segments were still identical would have derived, but whose segments were equipped with identical appendages. These would then have specialized, depending on their position, into antennae, jaws, or legs. But it is quite probable that things went altogether differently.

In almost all animals, in fact, the different positions along the animals' main body axis are far from equivalent to one another. The earthworm, for example, has no appendages, but its segments, even without considering the obvious decrease in circumference of those nearest the body's two extremities, are not all equal. At least during certain months of the year (those in which the animal is reproductively active), it is possible to recognize a certain number of segments that are slightly more swollen and of a different color than the others, which form the so-called clitellus. In this region the mucous material that will form the cocoons for the eggs and the nutrient matter (thanks to which the embryos will develop) is produced. The position of the clitellus can differ from one species to the next, but it is characteristic of each species; in other words, it occupies a well-defined series of segments. The location of the genital orifices (which I here use in the plural, since we are talking about hermaphroditic animals) is just as precise.

On the other hand a precise location of the individual structures along the main body axis does not require that it be subdivided into segments, given the precision with which the genital orifices are positioned—halfway down the body in the case of the genital orifice of female nematodes (cylindrical, often parasitic worms, e.g. the ascarids and the filariae); however, the male's orifice usually ends, together with the anal orifice, in a sort of cloaca located not far from the posterior extremity. In nematodes, as in earthworms, "information" exists that specifies and fixates the positions of particular organs with the same precision as, in an insect or a crayfish, the different positions are fixated for the development of the antennae, the mandibles, the jaws, the legs, and possibly other types of appendages.

We are therefore confronting a mechanism that allows one to specify positions along the animal's main axis. This seems to be a universal property of animals, one much more extensively shared than the presence of appendages. It therefore seems that one can legitimately state that, from the moment different animal lineages like arthropods or vertebrates "invented" their appendages, these grew at different points along the body's main axis, points that were *already* different insofar as position is concerned, even at a time when appendages had not yet appeared. If things really occurred in this sequential order, some difference between pairs of appendages must have existed since the beginning. Consequently it is probable that an animal with identical pairs of appendages along its main axis has never existed. If this is how events occurred, then asking which is the fundamental model for the appendages of arthropods (the antenna, the leg, or something else) obviously becomes a meaningless question.

However, this is not a negative outcome. It is, rather, an example of what can be gained in the understanding of living forms and their evolutionary change, if we do not limit ourselves to collecting and interpreting experimental data pertaining to the mechanisms of development, but integrate them in a perspective that takes the historical succession of evolutionary events into account. And there is another important aspect that must be underscored. In this play of structures that appear, disappear, or are exchanged for one another in specific positions, precisely these positions come to assume an autonomous existence of their own, almost as if they were chairs arranged around a table whose reciprocal relations (positions) do not change depending on the different identities of the people who sit in them. Is it therefore legitimate to say that these positions, *in and of themselves*, possess a concrete existence of their own?

The Zootype

In the last twenty-five years, molecular developmental genetics has produced an enormous mass of experimental data that lead one to answer this question in the affirmative. In *Drosophila* for example the formation of a pair of halteres is tied to the expression of a *Hox* gene named *Ultrabithorax*. It produces a protein that is never present in the anterior half of the animal, while it starts to appear beginning with the third and last segment of the thorax. In the absence of this protein, conditions in this segment come to be virtually identical to those in the immediately preceding segment, in which *Ultrabithorax* is never expressed. As a result, the second and third segment will form dorsal appendages (wings) of the same type and, more precisely, of the type that is usually produced by the second thoracic segment and not the third, which would instead form the halteres. Similar effects are produced by the missing (or erroneous) expression of other *Hox* genes in different positions along the animal's main body axis. These genes, like *Ultrabithorax*, can be considered responsible, in some fashion, for the location of nonequivalent positions along the *Drosophila*'s body. Many other discoveries made in recent years relate to this fact, which in itself is important; but it has also given rise to an equally large number of questions.

Two of these discoveries are of the greatest interest to us at this juncture. First, the different *Hox* genes that mark the different positions along *Drosophila*'s main body axis resemble each other closely, and it is legitimate to suppose that they derive from a common ancestral gene, by means of repeated duplications (a well-known phenomenon to researchers in molecular genetics) that cause slight differences in the different copies that have been created in this fashion. In any case their kinship is still recognizable.

Second, copies of the aforementioned genes are present in almost all animals (fig. 6), and basically in all these cases, the differential distribution of their products along the animal's main body axis results in the specification of distinctive positions, from which these

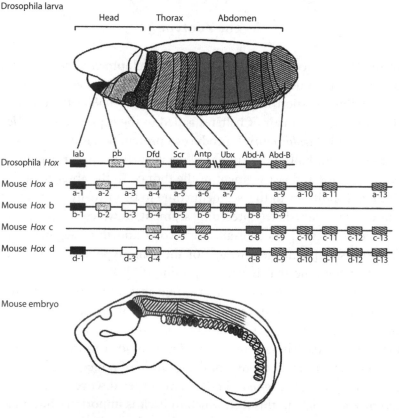

Figure 6. Zootypes.

animals will later develop characteristic structures. This discovery stimulated three English researchers (Slack, Graham, and Holland) to formulate one of the first great innovative concepts in the field of evolutionary developmental biology that was taking shape at the same time: the concept of zootype.

The *zootype* is the topographical plan according to which the different organs are distributed along the main body axis of all animals—or, to be more precise, along the main body axis of all animals with bilateral symmetry. At least as a first approximation in fact, it is more convenient to omit three or four animal groups in which it seems difficult to recognize an anteroposterior axis, similar to the kind we find in a vertebrate, an insect, a leech, or a nematode, for instance. We should remember that sponges also exist, a kind of animal whose body usually takes the shape of an irregular mass, either erect and branching, or encrusting, or massive—as do cnidarians, in other words jellyfish, corals, and hydras, which usually exhibit a radial symmetry around an axis (with the mouth at one of the two extremities) that does not necessarily correspond to the longitudinal axis of other animals. One should also consider the ctenophores or comb jellies, marine animals as diaphanous as jellyfish, and the little known placozoans, which are similar to little flat disks made of a few cells, but we will not discuss them further here.

All remaining animals form the great clade of the Bilateria, animals that exhibit bilateral symmetry, in which a front and a back, a top and a bottom are clearly recognizable. In any case these distinctions seem fairly clear in those Bilateria that are capable of movement. In these animals, the front is the extremity with which the animal always enters new locations, and the top is the side opposite the one the animal, if it moves on the ocean floor or on the ground, uses to remain in contact with the substrate. In these animals, therefore, we recognize a main body axis, which seems to be "the same" in all of them, and not only because in our descriptions we use an identical vocabulary (front, back, top, bottom, etc.), but precisely because, along this axis, different positions are marked by the borders of the areas in which the same *Hox* genes are expressed. All bilateral animals would therefore seem to have essentially the same set of *Hox* genes and in all bilateral animals these genes would be expressed in the same early phase of embryo development. Finally the proteins codified by these genes would be distributed along the anteroposterior axis of all these animals in a characteristic and basically unmodified fashion, thus defining basically invariable positions within all Bilateria. In other words it is precisely the presence of these genes and the typical spatial distribution of their products that form the basic plan for the animal's organization. And this is what has been called the zootype.

This molecular interpretation of the different positions along the body is characteristic of the evolution of many important concepts that have been used for (at least) a couple of centuries in the history of biology and are finally receiving their precise place in evolutionary developmental biology.

In Geoffroy Saint-Hilaire's comparative anatomy, the relative position of an organ was one of the most fundamental criteria used to establish its possible equivalence to the organ of another animal. But one and the same position can be occupied by organs that differ so much from each other, like the haltere and the wing of a fruit fly, that their equivalence tends to rest only on the identical position they occupy. Modern experimental biology, with its various "monsters," seems to widen the gap between the position and the specific qualities of the structure that occupies it. It suggests that both components, while ultimately shaped by different causal chains, combine to produce the results we observe.

Position and Structure

The order in which the mouth, genital, and anal apertures follow one another, in that specific sequence and not another, is something that is "written" in the animal itself. The three structures come to occupy positions that are precisely defined in the body's architecture, as designed by the *Hox* genes' products: positions that remain well defined even if one of these structures should be absent, or if others should substitute for them, each one obviously within its well-defined location.

To have discovered a molecular, in other words material, basis in the animal with which to explain an apparently abstract and conventional notion such as position is a very important achievement of modern biology. The result however is not conclusive, because the discovery of the existence of the *Hox* genes and the different modalities with which they express themselves leads one to a new series of important questions.

New problems become apparent both up and downstream. Upstream: what is it that determines and guarantees the specific spatial expression of each *Hox* gene? Downstream: in what manner do the products of the various *Hox* genes direct the production of specific structures in that precise part of the body? To answer these questions we must enter one of the most fascinating chapters of molecular biology, one that deals with the regulatory mechanisms involved in the genes' expression.

As mentioned, all animal cells possess identical inherited genetic information; they receive a copy of this information by way of a long series of cellular divisions that originates in a single initial cell, the fertilized egg. In each cell, however, only a portion of the genetic information is expressed; many genes remain silent. Regulating the expression of a gene means, above all, deciding if and when it should be transcribed. Other forms of regulation will subsequently intervene, for instance in the process that is often called the maturing of messenger RNA, in which the raw product that results from transcription does not necessarily coincide with the RNA molecule that will then provide the information for the synthesis of a protein. But we can limit ourselves to considering the process of controlling transcription.

The proteins encoded by the *Hox* genes are transcription factors, in other words molecules capable of interacting with other genes, allowing them to be copied or not, so as to form corresponding molecules of messenger RNA. This is why *Hox* genes are so important in specifying different positions along an animal's main body axis. Depending on which *Hox* gene is expressed locally (in some cases it is a question of a combination of several *Hox* genes), different non-*Hox* genes will be allowed to express themselves. And these different expressions can be crucial to the development of a specific organ, for instance for the realization of a haltere rather than a membranous wing. We should not be surprised, then, if the *Hox* genes are credited with a fundamental role in the origin and evolution of the structural plan for bilateral animals. If we admit the possibility of establishing a hierarchy organized according to the importance and function of different subsets of an animal's inherited genetic information, *Hox* genes would surely be near the top of such a hierarchy. They share a privileged position with other inhabitants in a very select Olympus; we shall meet some of these other inhabitants shortly, even though, in the next two chapters, we shall keep our distance from the account given by molecular developmental genetics and, with critical caution, face a problem that is all too easy to skip: if and to what extent it is legitimate to speak of genes as repositories of a program whose instructions, step by step, guide development.

"Master Control Genes"

Recent progress in molecular genetics and genomics has left us suddenly facing a series of unexpected discoveries that forces us to reconsider many concepts that seemed already well established.

A first set of data has been derived from what is colloquially designated as the sequencing of the entire genome of a growing number of organisms, including human beings, and from the ensuing interpretation of this information on the basis of the number of genes the organisms contained. Scientists already suspected that the differences between a fruit fly and a human being were not that great, if expressed in terms of the number of genes present in the nucleus of each cell in both species. But few would have bet that the number of genes in a human cell (some 26,000) was less than double those present in the cell of a *Drosophila*. And perhaps nobody expected the genes in a *Drosophila* to be fewer than those of a "simple" cylindrical worm such as *Caenorhabditis elegans*. And yet, that is how things stand. We also must resign ourselves to the fact that many other living beings, both plants and animals, have a more richly endowed genome than ours. Luckily, someone will say, the complex organization of the anatomical structure we are most proud of, which is to say, our brain, is determined by genes only in its general outlines; the rest is due to individual experiences: interactions with the external environment and, above all, with the social environment.

The question probably still spontaneously arises regarding the manner in which twenty or thirty thousand genes can account for the great diversity of cellular types that can be found in some zoological groups, starting with the vertebrates, and the complexity of anatomical forms and structures that the organized spatial distribution of these different cellular types allows for. For example, how many genes might be involved in giving a hand its shape? And how many genes might be involved in the building of the heart or the brain?

It looks as if, in order to guarantee the regular development of the peripheral nervous system of a *Drosophila,* it is necessary for the correct expression of about seventy genes to take place, while the molecules that mediate the interactions between the neurons of a vertebrate, or between the neurons and the matrix on which they extend, run into the hundreds. But these already quite respectable numbers pale by comparison with evidence that tells us that an animal's nervous system alone may account for the expression of up to half of its entire genome. A single organ endowed with a complex structure can involve the expression of a great number of genes: in the case of a *Drosophila*'s eye the estimate is about 2,500, which would correspond to about 18 percent of the total number present in the insect's genome. Naturally, of the many genes that may have a role in the realization of an organ, only a small fraction are expressed solely in this organ, and it is far from certain that these "exclusive" genes are the most essential to the process that leads to the realization of that organ.

It is not easy however to establish which genes are the most important in the construction of an eye, a heart, or a leg, even though some experimental data do occasionally seem to give us some precise information. In fact often the failed realization of an organ can be traced back to a mutation, or to the experimental deactivation of a specific gene. For example we know of some genes that are so important to the realization of the eyes of a mouse or a *Drosophila* that, in the absence of the proteins they encode, the animal is deprived of its eyes.

The real importance of these genes can be proven even more clearly if we modify the background against which their expression is realized—in other words, if we verify the effect a mutation of these genes has on individuals that also differ from one another in other genes, genes that we know somehow affect the realization of the same structure (in our case, the eye). Research of this kind has, since the early 1990s, led to the creation of the concept of *master control gene.*

A *master control gene* is a gene whose expression initiates the expression, following a well-ordered sequence, of numerous other genes whose products are indispensable to the realization of a specific organ. A mutation at the level of *master control gene* would therefore immediately halt what would otherwise be a long cascade of effects that would ultimately result in the construction of the organ. Naturally even a mutation that affects a gene "downstream" from the *master control gene* could have visible negative effects on the results of the entire morphogenetic process. But it could just as well initiate and continue to unfold up to a point of interruption caused by the failed expression of the mutant gene, while in the meantime allowing for the activation of other genes that are important to the regular developmental sequence and, perhaps, to the opening of a second, usually nonutilized, path, one that also leads to the construction of the relevant organ.

One such *master control gene* is, for instance, the *Pax6* gene, whose expression is necessary to the formation of the eye in animals as different as vertebrates, mollusks, and insects. Another example is *tinman*, a gene that occupies a position high up in the hierarchy of genes that are involved in the realization of the heart—in other words of that contracting, pulsating, section of the circulatory system in animals that belong to very different phyla.

Matters however are not so simple. The set of genes whose expression is necessary for the realization of a specific organ probably does not have the simple and rigorous hierarchical structure that the concept of a *master control gene* presupposes. At the very least, even if there are no really democratic relations between these genes, there is also no absolute sovereign. They are instead ruled by an oligarchy from which no true leader emerges. In the case of the genes involved in the formation of the *Drosophila*'s eye, for example, there are at least seven of them, to each of which the role of *master control gene* has been attributed by one developmental geneticist or another in the course of the years that elapsed between the end of the last century and the beginning of the new. Finally, however, none of these candidates passed all the tests and—the regard in which the concept is held by many biologists notwithstanding—the concept of *master control gene* now seems to be slowly losing its credibility.

Networks

One still must recognize that it was this hunt for the gene that was key to each specific morphogenetic process that brought the complex interrelations between genes (or between their respective products) to light, relations that more closely resemble a network than a linear hierarchy.

To understand something about the situation, we must appeal to a new scientific discipline, appropriately called the science of networks, in which mathematics, logic, information science, and electronics converge. This science is proving to be incredibly useful in analyzing the most diverse aspects of our existence: not only the biological, but also the social and economic.

Today we live in a web of networks, from those that concern our interpersonal relations (in the family, at work, during leisure time, etc.); to those in our road, rail, or air transportation; to those in our written, phone, and computer communications. The Internet's explosion has naturally been the most important stimulus—though not the only one—for undertaking a systematic study of these networks: how they develop and function, what their weak points are, and how they may change over time.

In the field of biology, models of this type have been successfully applied to the study of neural networks and to the study of the relations between species within an ecosystem. Of greater interest to us, because of the topics dealt with in this book, are the relations between different genes in the same organism. Of course these are indirect relations, mediated by their products, in other words by the proteins that a cell is capable of producing utilizing the information contained in each gene. The product A of a gene *a* can influence the expression of genes *b*, *c*, and *d*. In its turn the protein B, which constitutes the product of the expression of gene *b*, can influence the expressions of genes *d*, *e*, *f*, and *g*. And so on.

Leaving aside for the moment the nature of the specific elements (*nodes*) that make up a network and the types of relations that exist among these, all networks exhibit a set of properties and behaviors that are solely dependent on the number of nodes and the number and distribution of links that exist between one node and the other. More specifically the ability of a network to maintain its structure despite the interferences that may damage this or that node depends on the manner in which these links are distributed. Generally, most nodes have very few direct links with other nodes, but some are more solidly connected and a small number (sometimes only one node) are directly connected to many other nodes. An example of this situation can be seen when consulting a map of flight connections in a country in which this means of transportation is frequently used. There are many airports, most of which are connected to only a few other destinations, whereas there are a small number of airports with flights to and from a very high number of destinations. Thanks to these principal nodes, which are colloquially referred to as hubs, it is usually possible to almost maintain all links within the network as a whole, even if a couple of the minor airports are temporarily closed. And on average, this is true regardless of which airports are specifically affected. This ability of the network of air links to function even in the presence of small local malfunctions is, in technical jargon, its robustness—an important property, which naturally carries a price: the risk that the entire network will crash if the hub is the node affected.

Back to morphogenesis: if the control exercised by genes on the realization of the forms of living organisms were to follow a simple hierarchical cascade of ripple effects, one could hypothesize the

existence of some very simple evolutionary mechanisms. The mutation of a gene placed in a specific hierarchical position would have
no effect on the "upstream" portion of the hierarchy, which would
remain intact, whereas it would have foreseeable effects "downstream," all the less dramatic the lower the gene's position in the hierarchy. If, however, the relations between genes and their products
are not hierarchical, but reticular in nature, then these hypotheses
are not valid. And this has important consequences both for the individual development of the organism in which such a mutation
would occur, and for the possible future evolution of its descendants,
if we grant that the consequences of the mutation itself are not in-
compatible with survival. But these new perspectives offered by the
science of networks are, for the moment, only avenues that still await
exploration.

PART TWO

Constructing Form

Chapter 5

Evolution and Development

Genes and Determinism

From generation to generation living beings transmit copies of their genetic information to their offspring. When they reproduce by means of a nonsexual mechanism, for instance by producing buds that are destined to detach from the parents' body to give life to new individuals, the inherited genetic information remains unchanged, except for the effect of mutations, whose influence is usually negligible in the short term. When instead reproduction is intertwined with mechanisms of genetic exchange that are characteristic of sexuality, each new individual also gives rise to a new genetic combination.

Each new individual, therefore, represents a sort of small natural experiment. Generally we cannot predict how it will develop or its probabilities for success in the daily struggle for survival with the same precision with which we can formulate a similar prediction concerning genetically identical individuals. And this helps reinforce the widespread conviction of genes' quasi omnipotence. Each animal, each plant, each microorganism is what it is because it has precise inherited genetic information that determines its functional and structural characteristics, the way in which it develops (in multicellular organisms, in other words in those in which it makes some sense to talk about development), and also its behavior, the manner in which it responds to environmental stimuli.

Naturally one cannot state that everything genes accomplish is in the animal's, plant's or microorganism's best interest, given the genetic information it has received. No one will dare state that this is the best of all possible worlds. The embryo saddled by a malformation that arrests its development too early will not, and nor will those antelopes and zebras who fall victim to lions' aggressions because nature has not endowed them with the same olfactory sensibility and agile limbs that other antelopes and zebras have.

In any case, however, the earth does not fill up with monsters or weak and incapable creatures. Natural selection inexorably takes care of the situation—even if sometimes it takes awhile, and, perhaps, some unhappy creature on whose success no one would have bet even a dollar manages to avoid the rigors of natural selection for a long enough period of time to allow it to experience different environmental conditions in which, ultimately, its improbable endowment proves to be a winning combination. And perhaps it is precisely along this tortuous and risky path that a novel form destined to success with the passing of generations takes shape. The success of snakes, for instance, demonstrates how a terrestrial vertebrate without legs is not necessarily a cripple destined to disappear from the face of the earth without possibility of appeal.

This simple and yet grandiose picture of the evolution of living beings is based on two solid premises—on the one hand the existence of an inheritable genetic program, on which the characteristics of all organisms depend, both those of each single individual and those it has in common with other individuals of the same species, and on the other hand the efficiency of natural selection, which eliminates those variants that are least well adapted and favors the transmission of those genes and gene combinations responsible for the production of the most successful individuals to the following generations. Mutations and sexuality, together with changes in environmental conditions, as slow or fast as they may be, will be responsible for the production of new forms (the *modifications* that accompany the *common descent* of living beings, as acknowledged by Darwin).

This summary of biological evolution as given by school textbooks is simple and clear. And it goes without saying that we are referring to those that are up to date, in which the inherited genetic information of organisms is no longer described in the formal language that was once prevalent (when one only used the term *gene*, introduced by Wilhelm Ludvig Johannsen in 1909), but in the current language of molecules. What a pity, however, that this picture is only a crude caricature of reality.

Certainly it is easy for people to rebel against a view of life that appears to be dominated by an all-powerful genetic determinism. Is it possible, one may ask, that this also applies to human behavior? What is the fate of our free will? And are all the noblest expressions of our spirit just the inevitable realization of a program inscribed in our DNA? What can that know-it-all molecule possibly understand of the emotions a human being may feel when listening to a Beethoven symphony? And of the harmonious shapes that a sculptor derives from a block of rock; is it at all possible that these were pre-inscribed in the sculptor's inherited genetic information to a greater extent than they were in the block of marble before his chisel went to work?

And yet, one cannot deny that genes are important. Many successes in the fields of medicine, agriculture, and animal husbandry, tied to interventions that depended, in one way or the other, on the action of individual genes or their specific combinations, have demonstrated this experimentally. Even without knowing anything about what Mendel, Morgan, or Watson and Crick would have discovered, the Mesopotamian farmer who derived wheat with its abundant ears from the humble and not very productive grass of the steppes, had already acted in accordance with these discoveries. Analogously the efficiency of the process of selection that managed to derive the Doberman and the poodle, the greyhound and the Neapolitan mastiff from the wild wolf can bear witness to the genes' targeted action.

Problems arise when one attempts to attribute a greater importance to genes than they probably actually have. When one reads most of the biological literature of the last decades it really does look as if everything is decided by genes and as if the rest of living matter had no other role but that of executing orders that cannot be questioned, even if factual reality (in other words confrontation with the surrounding environment) sooner or later proves that by following those orders our poor organism has really gotten into a lot of trouble.

Perhaps, however, reality is different. Even though it is impossible to deny that genes have very important control functions in the development of the organism and the explication of its everyday activities, it would be prudent to distance ourselves from a view of living beings that is too exclusively (I was about to write: obtusely) focused on the gene. Such an adjustment of perspective is not important only from the point of view of developmental biology or physiology. It can also have a profound effect on the way in which we describe, analyze, and interpret evolutionary processes.

All too often the textbooks also offer us a short-sighted perspective that reduces evolution to variation over time, within a population, of the frequencies of specific genetic variants (alleles) or their combinations. On this basis, it is implied, one should be able to address all the great problems of evolutionary biology: adaptation to the environment, the development of new species, and even the origin of new structural plans, like the appearance of the first bird, or of the first plant with flowers.

Ultimately—it is either stated explicitly or implied—where is the problem? Genetic mutations and the phenomena associated with sexuality do not cease to produce new gene combinations to which structural novelties or innovations in the way organisms function correspond. In other words there is a continuous production of variability within natural populations, and this variability is somewhat like the block of marble in Michelangelo's hands: natural selection will, day after day, eliminate the excess, allowing for the emergence of the part that responds best to the adaptive requirements of the environment in which it finds itself. And since mutations and sexuality continue to produce new variants, while the environment continues to change, the product of natural selection will be a population in a state of continuous change, up to that point when some unfavorable circumstance will lead it to extinction. Other circumstances will instead favor the division of the original population into two "offspring" populations that, with the passage of time, may diverge to the point of becoming two different species. This is an important argument in the context of evolutionary biology: we shall not concern ourselves with it further in these pages.

Possible Butterflies, Real Butterflies

In fact the problem I am most interested in is another—that of the nature and origin of that variability on which natural selection can operate. It is not sufficient to know the mesh size of a sieve to make forecasts about the characteristics of the flour or sand that will pass through it. It is also necessary to know what mix of materials we place in it. And one cannot tell whether in this material there are particles of all possible dimensions. Rather it is probable that the history of this material (where it is from, how it has been collected, and what treatments it has undergone before being placed in the sieve) has left its mark. If this is the case, then it is probable that some fraction of small particles that would have easily passed through the sieve, but which in actuality did not, is missing from this material because it was not present in the material to be sieved in the first place.

Certainly, one will claim, the Earth has finite dimensions, and therefore there is no room on its surface for all the hypothetical living organisms that would be capable of living on it. It is easy to invent butterflies with wings decorated with patterns that differ from all those known to us, butterflies that might perhaps have excellent chances of survival but which, to put it simply, nature has never experimented with. Or if it did so, we don't know about it because the probability of a butterfly being preserved as a fossil is close to zero. Or perhaps nature will conduct such experiments in the future, only we have no way of forecasting if and when it is probable this might take place.

But the potential diversity that could enrich the already extraordinary chromatic diversity of butterflies does not in itself constitute a problem. We should not worry that the universe of possible butterflies is greater than that of real, existing, or past butterflies. We should really be asking ourselves a different set of questions. Among the many, perhaps we would do best to start with this: is there a precise border between possible and impossible butterflies? Are there butterflies that I could draw on paper, but will certainly never find in nature?

The answer may seem easy. Two hundred years ago there were still areas of the Earth that had never been reached by any traveler, and it was possible to imagine that there, in addition to the *leones* that (on almost all maps) inhabited almost all the unexplored areas, there could also be butterflies with phantasmagorical colors and patterns never before seen. And quite frequently these expectations were met. Today, however, we live in an era when the entire Earth has been explored and such remote corners are close to disappearing.

But it is not by enumerating all known species that we shall be able to exhaust our knowledge of all possible models and thus discover the border that separates them from those nature will never be able to generate. As I said, the universe of possible forms can include a large, perhaps infinite, number that have not yet been realized. How then will we ever be able to know if one of our virtual butterflies can be included or excluded from the list of possible butterflies?

It might seem insolent presumption toward nature to claim to be able to know what it can and cannot do. At the very least there is the risk, sooner or later, of being proven brutally wrong by the facts. But living organisms are not created and undone on a whim. Centaurs and chimeras, in whose bodies parts of different creatures are united, remain confined to the world of the fantastic. In the real world we instead find calves with two heads and *Drosophila*s with a pair of legs in place of their antennae. In the case of the calves, the two heads, whether complete or not, are identical to each other (and are, therefore, legitimate calf heads), and their position is not casual, but conforms to the animal's overall symmetry. In the case of the *Drosophila*s, the supernumerary legs that have taken the place of the antennae are still *Drosophila* legs, down to the last hair.

"Monsters" also obey some laws, just as "normal" animals do. As a matter of fact—and this has opened the doors to extraordinary possibilities of experimental inquiry for the biological sciences—the laws that "monsters" obey are the same as those that govern the development of normal individuals. And they are, in both cases, laws. We should take note of the double scandal: calves with two heads and *Drosophila*s with four wings would seem to be impossible creatures, and yet nature is able to produce them. Scolopendras with twenty-two pairs of legs would seem a banal variation on the more common ones with twenty-one pairs, but nature is incapable of producing them. This then is the direction in which we can focus our attention—on the inquiry regarding the border between possible and impossible forms; on those laws or rules whose existence we begin to suspect when our expectations are so blatantly proven wrong.

Evo-devo

In the third chapter we talked about forbidden, allowed, and preferred numbers. All these stories lead to an identical conclusion, simple but important. It appears clear, in fact, that of all the imaginable living organisms, those that actually occur in nature satisfy the following two conditions: first, it is possible to construct them; second, they function well enough to allow them to perpetuate themselves, at least for several generations. Simple common sense, some will say. Perhaps, but it is a common sense that seems to have been absent among biologists while they remained divided in two groups, the developmental biologists and the evolutionary biologists, with little interest in a shared dialogue. Both formulated apparently identical questions that, however, in each of the two worlds came to assume different meanings and gave rise to research programs that were not compatible.

When experts in evolutionary biology ask themselves why does this butterfly have four wings and why do its wings have these particular patterns, which are different from those of other butterflies, their questions are set in the context of evolutionary history and adaptation to the environment. Our experts, for instance, will try to reconstruct the structural plan of the first butterfly and formulate a hypothesis about its wings: Were there four of them at that time? If yes, were there four because the first butterfly inherited them from its immediate ancestors, or was it instead an invention of the butterflies, which derived from ancestors without wings or with a different number of wings?

Insofar as the colors and the patterns of the wings are concerned, to project these minute details on a hypothetical primordial butterfly that lived in a distant era can be a somewhat risky undertaking, but our expert in evolutionary biology has many sensible questions to ask about the colors and patterns of an actual butterfly. What advantage can it derive, for example, from having a wing with red, blue, and yellow patterns above, while below black and various shades of brown predominate? Is there a connection between the colors and patterns and the butterfly's social life, in relation to its own kind or in relation to possible predators? In an attempt to answer these questions, it will be useful to compare the results of patient observation in the field with that obtained in the laboratory, for instance by exposing butterflies with different patterns (natural or the result of targeted manipulation) to the attention of possible predators, in other words of a kind that they might encounter in their natural environment. Or, perhaps even better, to the attention of those predators that their immediate ancestors most probably had to face.

For developmental biologists, instead, to ask themselves why a butterfly has four wings and why they are covered with certain patterns means putting on their lab coats and researching, at a cellular or molecular level, the developmental mechanisms responsible for sooner or later transforming a wormlike caterpillar without wings into an adult with wings. Or to research those molecular mechanisms responsible for the formation of wings only starting with the second or third of the three segments of the thorax, but not the first, or any of the abdominal segments. Finally, it means asking oneself when and how, and under which genes' control, a set of patterns will be traced on the rudiments of future wings that will finally become visible only with the emergence of the adult.

The research objectives of the evolutionary and the developmental biologist are therefore very different, as are their research methods and the journals in which they usually publish their findings. The conferences they attend are also distinct, as are the paths their academic careers take. But one should not criticize either group excessively for their being to some degree impervious to what is occurring in the other field. The vastness of the horizons of modern scientific research, which is increasing with each passing day, precludes a detailed knowledge even just outside the narrow range of problems, objects, and study techniques that are ultimately the object of each expert's specialization. For developmental biologists who have always worked with *Drosophila*s, just dealing with the developmental biology of vertebrates may be difficult. Then imagine how distant the problem set of evolutionary biologists must seem to them. And vice versa.

And yet stories like that of the giraffe and the scolopendras, with their numbers sometimes set in stone and sometimes impossible, demonstrate that the dialogue between evolutionary and developmental biology is necessary if we want to understand why some forms occur in nature while others, apparently very close to them, do not at all.

For some years now, these two branches of biology have begun, finally, to meet. From this encounter a new discipline has arisen that has been given the name evolutionary developmental biology, derived from the title of a seminal book by Brian Hall, the first edition of which appeared in 1993. Soon the growing interest for these problems created the need to find a colloquial expression that was shorter and easier than the correct but academic term *evolutionary developmental biology*. It is for this reason that today one generally refers to this discipline simply as *evo-devo*.

Chapter 6

The Logic of Development

From Mechanics to Molecules

If we draw up a balance sheet of these initial years of evo-devo biology it is fairly easy to note that the two "mother" disciplines have contributed to it in ways that differ significantly in terms of importance and visibility and not only insofar as the content of ideas, methods, and results are concerned. Developmental biology has definitely dominated this phase, thanks to the spectacular results of what we can call molecular developmental biology.

For at least three-quarters of a century, starting with the rediscovery of Mendel's laws (they had already been formulated by the Bohemian abbot in 1866) at the beginning of the twentieth century, the science of heredity has concerned itself almost exclusively with the transmission of hereditary factors from one generation to the next. Developmental biology, in other words the study of those mechanisms by means of which genes exert their influence on those processes that lead to the realization of an animal or a plant, instead had difficulty taking off, more because of technical difficulties than for theoretical reasons. The cutting edge of developmental biology was therefore represented by research in experimental embryology, for instance the isolation of blastomeres (the first large cells into which the fertilized eggs divides at the beginning of embryonic development), transplants, and grafts. These experiments were often decisive in introducing fundamental concepts such as organizer, competence, and induction into biology, and even today, they generate a sense of respectful admiration for the ingeniousness of the projects and to an even greater extent for the manual dexterity that was necessary to complete them.

Compared to the best aspects of this "mechanical embryology," many contemporary experiments in developmental biology, in which an embryo is subjected to the influence of some chemical agent, like cannonballs being aimed at targets too delicate to be hit, as current jargon would have it, in a "surgical fashion," may appear fairly coarse.

Today, instead, we have at our disposal an entire arsenal of molecular techniques that allows us to engage in a selective fashion with a specific process, on a specific organule, and on a specific phase of development. This state of affairs has been made possible by the extraordinary progress, both theoretical and technical, that has allowed for the identification of a large number of genes involved in important phases of development. In the case of these genes, it has also allowed for the identification of the manner and the time period in which they express themselves, their reciprocal influence, and finally also their results. This is why this modern "molecular embryology" is just another aspect of that developmental genetics that has taken such a long time to evolve.

Traditional experiments in crossbreeding, for example those undertaken on *Drosophila* almost one hundred years ago in Thomas Hunt Morgan's famous "Fly Room" at New York's Columbia University, could lead to recognizing the existence of hereditary factors capable of influencing a certain phase of development, but that is as far as they could go. Certainly it was already useful, also for medical purposes, to know that there exists a gene whose defective functioning, in case of a mutation, causes achondroplastic dwarfism. This is a gene that contains the information for the synthesis of a protein that is indispensable for the formation of the cartilage around which long bones are formed. But there is still a long way to go from this level to knowing the actual role of a gene in a developmental process. A long way especially when one does not dispose of anomalies capable of casting a glimmer of light that will aid our understanding of the normal course of development.

The Lawfulness of "Monsters"

I should add, however, that beginning in the first half of the nineteenth century, one had tentatively looked at "monsters" as a possible source of precious information on developmental mechanisms. This was a perspective that bore witness both to a specific will to look at the study of development scientifically and to distance oneself from the irrational attitude with which for centuries the topic of monsters, and not only human ones, had been addressed. It was an argument often enveloped in sulphurous vapors or catalogued among the omens that could foretell disasters.

But in the years 1832–37 Isidore Geoffroy Saint-Hilaire, son of the great zoologist and anatomist Étienne, whom we have already had occasion to discuss, published a treatise on human and animal teratology in which the anomalies were presented according to a precise classification, with an almost Linnaean flavor that suggested the existence of some fundamental set of laws, to which even the most aberrant teratological cases could be referred. In other words, this was a rationality that allowed one to glimpse the consequences of an alteration in the normal course of development. But such alterations could occur at as many different points as there were recognizable classes of anomaly in the "monsters" themselves.

The simple existence of "monsters" therefore becomes a sign of the possibility of analyzing the development of an animal as an organized sequence of events, each of which is likely influenced not only by the regular development of the preceding phases, but also by the influence of some new factors. Later one would say: the influence of a new gene that had not been expressed up to that point in development. The mutation of such a gene could deprive the embryo of a necessary factor for the regular unfolding of events according to their usual phases. The resulting consequence could be either an arrest in development or the unfolding of an alternative course of events. Different anomalies in the same animal species could be related to mutations in different genes whose normal expression, necessary for the unfolding of a specific morphogenetic process, could have taken place at different periods of embryonic development.

In the last quarter of a century, these suspicions have become certainty with the growing availability of increasingly powerful and precise methods of inquiry that have led to the identification of a large number of genes that are active in the different phases of morphogenesis.

Initially, however, this progress in developmental molecular genetics, even though resounding and of great importance for all of biology (it finally made the connection between developmental biology and the science of heredity very concrete), provoked little interest among experts in evolutionary biology—due to the fact that developmental molecular genetics was initially limited to only a few model species, chosen on the simple basis of opportunity and cost. These were also supposed to be species easy to raise and manipulate, and from which it would be possible to obtain a large number of eggs and embryos. Preference was therefore accorded to species that were already available in laboratories, and the initial choice, in favor of the fruit fly, the famous *Drosophila melanogaster*, was almost obligatory. *Drosophila* thus became highly popular among the organisms that biologists used as models, repeating and to some degree even overtaking its popularity during the period between the world wars, when this little insect was the most widespread and best known of all the animals used in genetics laboratories. Toward the end of the 1980s, however, a change took place that would profoundly mark the ensuing development of these studies, leading them to an inevitable encounter with evolutionary biology.

In fact the hunt for genes that are expressed in the first phases of development and are therefore, in all probability, involved in the development of the embryo's structural plan, was progressively extended to species other than *Drosophila melanogaster*. As could have been expected, these studies soon involved a vertebrate. The animal in question was a *Xenopus*, a sort of African toad with a smooth skin, which had already been a habitual guest in many labs devoted to developmental biology. Later the chicken and mouse were studied; as a mammal, the mouse represented a more plausible surrogate for the animal whose biology researchers were especially interested in—the human being.

A Worm Enters the Scene

Meanwhile, however, experimental research acquired a new model, destined to compete with *Drosophila* in the hit parade of the most studied animal species. This rising star was a minuscule roundworm, whose scientific name, *Caenorhabditis elegans*, is very often reduced, colloquially, to the simple "*C. elegans*."

The reason to place this species next to *Drosophila* or the mouse, among the restricted number of model species systematically used in research, is above all a property that the old zoologists at the end of the nineteenth century discovered in some groups of invertebrates, which included nematodes or cylindrical worms, and to which *Caenorhabditis elegans* belongs. This property, known as eutelia, refers to the number of its cells being invariable; in other words, all the individuals of a certain species (and of the same sex) have an identical number of cells, and this number is divided among the animal's organs according to proportions that are themselves invariable. For example in all male *C. elegans*, the nervous system is formed by an identical number of cells, and the same goes for the digestive tract, the excretory system, and so forth. Since all of an individual's cells are derived by successive divisions from a common ancestor (the fertilized egg), the numerical invariance of the products of these repeated divisions and their rigorous distribution between the individual organs led to the suspicion that the series of cellular divisions producing all the animal's cells were the same for all individuals in the species. It would therefore have been interesting to reconstruct this sort of cellular genealogy and then attempt to determine the different moments starting from which an individual cell's destiny and that of its potential descendants are definitely committed to a given cellular specialization.

Naturally it is only reasonable to attempt such a reconstruction if the eutelic creature, in addition to being easy to raise, is also formed by a relatively modest number of cells. Otherwise it would only be possible to follow the first divisions that occur after the egg is fertilized, divisions that give rise to an increasing number of blastomeres. For a while, these blastomeres all appear at the surface of the embryo, but quite soon an external layer of cells will cover the remaining ones, making it difficult to watch their behavior.

In the case of the diminutive *C. elegans*, all the conditions were in place to make it an excellent model organism. It was easy to raise, and the number of cells that formed the body of an adult (fig. 7) was modest (about one thousand total). It was therefore worthwhile pursuing this venture.

The complete reconstruction of the cellular genealogy that is displayed during the embryonic development of *C. elegans* was ready in 1983. This was undoubtedly a very important result, even though (or perhaps, precisely because) the picture that emerged was very different from what was expected, that all the cells of a same type, involved in the formation of a specific organ, must derive from the same precursor (or the same couple of precursors, one to the right and one to the left of the animal's symmetry plane). Reading this same story from a different perspective—cells that were related to one another, in a genealogy that descended from the fertilized egg, should possess similar characteristics and be found within the same organ. But in actual fact the situation was quite different, for it was discovered that the degree of kinship between any two cells in this genealogical tree did not predict whether they would adopt a similar specialization. It

Figure 7. *Caenorhabditis elegans.*

almost seemed as if the cells that made up each organ were recruited by chance, within the available cell population, the only criterion necessary being the total number of cells constituting the organ itself. But his new hypothesis did not fit the fact that, while apparently devoid of any logic, the recruitment of the cells to form different organs was rigorously identical in the different individuals studied.

It took some time to understand that, faced with this apparent paradox, researchers would have to observe development in three dimensions, in other words in the manner in which it actually takes place, instead of adopting the conventional representation of cellular divisions by means of a two-dimensional graphic that only shows filiation relations (fig. 8). The fact is that each cell occupies a specific position in space and therefore has specific relations of contact, or vicinity, with other cells.

Given the precision with which cellular divisions occur in the course of the embryonic development of *C. elegans*, those which the cell comes into contact with are presumably the same in all embryos, but are not necessarily, or always, those it is most closely related to

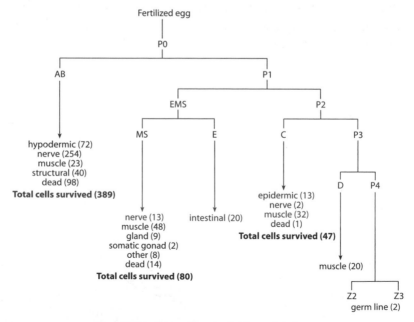

Figure 8. The first embryonic divisions of *C. elegans*.

genealogically. These circumstances become clearer if we keep in mind the appearance and size of the diminutive worm, with its cylindrical, narrow, and elongated shape, and also remember that it is constituted of a number of cells that only at the end of the developmental process reach approximately the one thousand mark. For developmental biologists, this entailed a revision of what until that moment they thought was known about nematodes.

A Mosaic, or Perhaps Not

In fact the textbook notion was that the embryonic development of these little worms was of the "mosaic" variety, which was juxtaposed to the "regulative" development typical of the sea urchin or the frog, for instance. A mosaic type of development means that very early on, already after the first cellular divisions, each blastomere acquires a specific value; in other words, it specializes in a specific direction, which means that as a consequence many alternative paths of development are precluded to its descendants. And this specialization becomes more focused as the divisions proceed, the number of cells increases, and the first rudiments of the future body architecture begin to appear in the embryo, which up to then had the appearance of a simple mass of cells more or less similar to one another.

Because of this precocious specialization, if we eliminate one of the blastomeres, the embryo will never be able to form those parts that grow from cells descended from the blastomere that has been eliminated. In embryos of a regulative type, however, the potential of individual blastomeres remains more open, and so the elimination of one can be compensated for by the redistribution of the cells derived from the remaining blastomeres, which in any case provide the material for all the animal's organs.

What we now know about *C. elegans* leads us to reevaluate the contrast between these two types of embryonic development (mosaic and regulative), a contrast whose importance had previously been exaggerated. It seems that *C. elegans*'s blastomeres are not as autonomous in their decisions as was previously thought. Their destiny and that of their offspring is instead conditioned, as in the case typical of regulative development, by the reciprocal influences that blastomeres exert on one another.

One Model, Several Models

Caenorhabditis elegans is one of the new model species that has entered biology laboratories in recent decades, joining such traditional guests as the mouse and *Drosophila*.

Among the animals in this group, at least one other must be noted—the small zebra fish, one of the most frequent inhabitants of the first small aquariums of tropical fish, which now, above all because of the excellent transparency of its embryos, has become one of the principal research subjects of developmental biology. In a short period of time the number of these model species has become greater; these species have become the protagonists of recent strides in developmental molecular genetics and the comparative spirit that for a long time had seemed to be lost in many areas of biology seems to have been given new life.

In developmental biology laboratories in particular, the research into general principles, into mechanisms common to all species, was dominant. From everything that was learned from experimenting on *Drosophila*, what really became a matter of interest was not, in and of itself, the expression of a specific gene at a specific phase of the fruitfly's development, but rather what could be extrapolated from this knowledge and generalized to other animals, humans included. Naturally, precisely in order to see to what extent an experimental result pertaining to *Drosophila* could be generalized, some comparisons with other species became necessary. And this is where, almost suddenly, the doors opened to evolutionary biology. In fact, comparing the results of studies undertaken on different animals by developmental geneticists, it was possible to highlight both the similarities and the differences between species at the level of the genes involved and in regard to the patterns and consequences of their expression. On the basis of these comparisons, developmental biologists could concretize their generalizations about the genetic and molecular mechanisms common to the most diverse zoological groups.

For the first time evolutionary biologists opened a window onto the developmental mechanisms and the genes involved in this process. They thus began to ask to what extent the structural similarities or differences between species belonging to different evolutionary trees could be related to the either conservative or innovative character of the genetic and molecular developmental mechanisms involved. This was the first strong encroachment on the century-old barrier separating the two disciplines. In other words, the time was ripe for the rise of an evolutionary developmental biology.

Born, therefore, as a result of the first comparative studies on the genes involved in development and their expression, evo-devo has been characterized up to this point by a strong molecular-genetic imprint. Put another way, in the dialogue between developmental and evolutionary biology, the first of these disciplines has spoken with a stronger voice in recent years. It would not be a bad idea, however, to give evolutionary biology the space it deserves, as I will attempt to do, to a certain degree, in the following chapters.

The Adult's Questionable Uniqueness

Certainly in order for a dialogue to develop it is best if the two interlocutors are placed, to the greatest extent possible, on the same level. And this cannot certainly be taken for granted in the case of the two partners in the evo-devo dyad. I am not alluding here to the differences in research methods, or to the fact that developmental and evolutionary biologists have, until recently, formed two separate scientific communities, facts previously mentioned. What is important, to enable dialogue, is the overall Weltanschauung, which is not really the same in the two research traditions. I believe that evolutionary biology has been able to alter its worldview more than developmental biology. This at least in one important way: its attitude regarding a finalistic view of the phenomena being studied which has become more critical on the evolutionary side.

In a scientific interpretation of evolutionary processes, there is no room for finalism. During the uninterrupted chain of generations, the first terrestrial vertebrate did not live with the goal of generating, at the appropriate time, a human being, or for that matter a mouse, a dolphin, or a tyrannosaurus. This does not mean that the evolutionary processes that have led to human beings, mice, dolphins, and tyrannosauruses happened by chance, outside all laws of nature. It does signify, however, that in the course of time the laws of nature have operated in a multitude of contingent situations, none of which had been previously planned.

Evolutionary history does not follow a plan, but lays out a pattern whose logic can only be interpreted after the fact. In practical terms this means that today those who dispute this view are those who abuse the term "biological evolution" in order to tell a story that does not belong in the purview of the sciences. The perspective of Intelligent Design, which allows for the development of living beings through time, but precisely only as the unfolding of an intelligent plan that precedes and guides history, for instance, belongs to such a group. But since this is not a scientific hypothesis, we do not have to discuss it here. We must instead deal with that form of creeping, undeclared finalism that survives in the way we currently understand development. In a certain sense it is a question of the survival of a naïve, pre-scientific view that privileges the adult as the "true," legitimate condition in which the characteristics of a species are fully manifested, while it reduces other phases of development, starting with the egg, to being simply preparatory phases. Their significance can therefore only be understood as a function of the adult toward which they "tend." Naturally what can be said about the egg can also be said about the seed, which, in this view, only has "value" insofar as it contains within itself, potentially, the complete plant.

Boxed Generations

It is precisely on the basis of these popular notions, moreover, that, beginning in the seventeenth century, the scientific study of the development of the animal and the plant is grafted. Today it is easy to smile about the notions of the preformationists according to whom the egg (if not already the minuscule spermatozoon) already contains a miniature copy of the future adult animal. It is easy to smile, especially when preformationism presents itself in its most extreme form, suggesting not only that in each egg there is a tiny, but complete, primordium of the future animal, but that the animal in its turn contains the even smaller primordia of future generations and so on. In other words we are faced with a progression of Chinese boxes that could make us ask if, in such a scenario, it would be more plausible to expect a finite sequence of preformed generations or instead an infinite sequence.

Naturally it is not easy to point to convincing observations in support of this hypothesis, even though Charles Bonnet's preformationism is perhaps comprehensible. Around the middle of the eighteenth century he observed a small aphid (green fly) being born to a non-impregnated mother: today we would classify this as an instance of viviparity associated with parthenogenesis. What is more, among aphids this is quite a normal occurrence, which in a number of species occurs for several successive generations. It is hard to believe the mother did not already have within her the primordia of the offspring she would later have given birth to, and, in actual fact, in female aphids these primordia may be present before they themselves have been born!

Already during the eighteenth century, however, this naïve preformationism, which presupposed the existence of the future adult (in other words of the next generation, at least) inside the egg or the spermatozoon, was challenged by a different theory, epigenesis, according to which the new organism would develop gradually, passing through a series of stages whose forms are not predetermined in the gametes. This was certainly a more sober alternative, which easily accorded with the data from observation. It was, however, a clearly incomplete hypothesis, incapable of explaining the great complexity of the structures that gradually unfold in the course of development and, above all, the close resemblance between parents and offspring.

This gave rise to the hunt for those minuscule material particles, whatever their appearance, size, and nature, that, being transmitted from one generation to the next, ensure the fidelity with which the characteristics of the parents reappear in their offspring. Before being materially recognized and localized, these particles would receive many names. They are, for instance, Charles Darwin's "gemmules" and August Weismann's "determinants." Naturally, they came ultimately to be called genes and referred to long strands of DNA chained together with other similar segments to form chromosomes. The premises for a future integration of the science of heredity (genetics) and developmental biology were finally present. It was an integration that took awhile to take shape, since it was rejected by some of the first geneticists, like Morgan, and by a considerable number of the creators of what in evolutionary biology is called the Modern Synthesis. When an integration of embryology and genetics was reached, however, the view that emerged was perhaps closer to preformationism than to epigenesis. This is the case, at least, in the version that became more established as gradually the structure of DNA was understood, and the manner in which it is transmitted with each cellular division, the role of protein synthesis and, therefore, the control of a vast array of cellular activities by means of these proteins became clearer. The era of molecular biology had begun. Not long after, the era of computers and computer programs also materialized. Machines and algorithms soon offered a pretext for the development of metaphors that were destined to have enormous success in biology.

According to Program

DNA thus becomes the repository of genetic information, and development is described in terms of the execution of a program. An influential embryologist like Lewis Wolpert can ask himself if, with more detailed knowledge of the genetic program inscribed in the nucleus of the egg cell, it might not be possible to "calculate" an embryo, in other words to foresee the sequence and unfolding of those processes that will lead to its transformation into an adult.

The situation is dangerous. On the one hand biology can celebrate the reaching of an important goal: the identification in the form of DNA of the material substrate for the transmission of hereditary information, a discovery that opens up unprecedented horizons to experiment (but also to the possibility of manipulating the genetic information of a plant or an animal, human beings included). On the other hand there is the risk of taking the genetic-program metaphor too seriously and, above all, of placing the gene on too exalted a pedestal. Most important it becomes exceedingly easy to adopt a deterministic view of development that is dangerously close to finalism.

In the egg, it is said, one finds the *program for* the construction of the adult animal. In the seed there is the *program for* constructing the plant with its full complement of branches and flowers. It doesn't matter who put the program in the egg or the seed; it is of little use to underscore the fact that there is no creator involved, and that the genetic program in the seed or the fertilized egg is, rather, the result of a puff of wind or a dialogue between gametes. The problem remains.

In my opinion, this view limits our ability to formulate intelligent questions about developmental mechanisms (and, above all, their origins) and makes the dialogue with evolutionary biology more difficult. This is an evolutionary biology that has basically rid itself of finalism and is therefore ready to take each organism, each generation, as the contingent and provisional result of a history that may perhaps continue, following rules that are known or can at least be investigated, but which do not follow any preordained design or program.

Cuticle, Cuticles

Let us take the example of the cuticle that envelops the bodies of numerous animals—nematodes and arthropods particularly. We have already come across nematodes: they are those worms with a cylindrical body, not divided into segments, to which the by now famous *Caenorhabditis elegans* belongs. Before this little worm became the focus of the news, the most famous nematodes were its parasitic relatives, the diminutive human pinworm, the voluminous ascarid, and the hair-thin filariae. Arthropods, those animals whose bodies are subdivided into segments, and carried by a varying number of articulated appendages, we know even better. In addition to centipedes with their incorrigible predilection for an uneven number of legs, insects, spiders, scorpions, crabs, and crayfish are all also arthropods. These are animals whose cuticle is often very thin and flexible, as in spiders and flies, whereas there are other animals whose cuticle is thicker and less flexible, such as scorpions and scarabs; and in some cases the cuticle is mineralized, as in crabs.

It is easy to verify what these animals do with their cuticle, which is, in fact, their skeleton. A skeleton that differs greatly from ours, not only because it is not made of bone, but above all because it is an external, not an internal, skeleton. In both cases, however, we are dealing with a more or less rigid series of elements on which the muscles can attach themselves, thus creating a useful system of levers. The sudden takeoff of a fly, a grasshopper jumping in the grass, or a crab that has decided to pinch one of the toes in our foot between its claws proves just how well this system functions. And there is no doubt that an arthropod's cuticle, especially when it is thick and robust, provides a protection to the little animal that a soft snail may well envy. This is also true for the cuticle of *Caenorhabditis elegans* and the other nematodes that, by not being subdivided into distinct "skeletal" pieces, can only offer much lesser mechanical advantages by comparison with that of the arthropods.

Although it is easy to understand what uses this cuticle serves the little animals today, it is not as easy to imagine how and why it came about. Let us try and see if developmental biology has some good suggestions to offer us.

In order to see an arthropod (or a nematode) covered by a cuticle, we don't have to wait for it to hatch from the egg. The first cuticle, in fact, forms around the embryo. Please note, however, that in this phase of development, there is not yet any muscular activity, and that, for the moment at least, the mechanical protection of the embryo is assured by the shell (which contains the embryo, but which is not glued to it like the cuticle, and which is produced by the cells that envelop the animal) more than by the cuticle itself. But there is more. Before the little arthropod or nematode, having completed embryonic development, exits the eggshell and begins an active life, it usually undergoes one or two molts. Exactly: the embryo undergoes one or two molts while it is still inside the eggshell! If the presence of an embryonic cuticle seems strange on its own, the molts it completes in this phase of development seem even stranger.

We are in fact used to considering molting as a necessary "change of armor" that the animal undergoes several times in the course of its growth until it reaches adulthood. But, in reality, things are not this simple. In the case of a butterfly's caterpillar for instance, the cuticle that forms the robust capsule enclosing the head, and the cuticle that covers the jaws and mandibles is rigid; these parts, therefore, cannot grow unless a molt intervenes. But the caterpillar's long and voluminous abdomen is covered by a much thinner and, to a certain degree, flexible cuticle, and, within this, room can be found for a mass of inner organs that, mouthful after mouthful, seem to grow as you watch. If we then shift our attention from arthropods to nematodes, we find that after the last of the four molts that mark the development of these worms subsequent to the hatching of the egg, the animal still hasn't reached its final size.

Thus, molts aid growth, though a molt is not always necessary for growth. But a molt inside the egg, what can its purpose possibly be? Perhaps, one could say, it is a question of precaution. It is possible that the first nematode, or the first arthropod, or some remote ancestor of theirs started to form its cuticle at the last minute, more or less at the time it was about to leave the egg. There was therefore no reason to be too complacent. In fact, there was the risk that an excessively thick and resistant cuticle, capable of protecting the animal, but above all of providing a solid point for its muscles to attach to, would not be ready when it was due to hatch, in other words when most necessary. Natural selection would therefore have favored individuals capable of producing the cuticle at a late stage of embryonic development. That way, at hatching time, it would already be available.

Continuing along this path, the production of the cuticle might perhaps have been brought further forward so as to allow for one or two embryonic molts. But this is precisely the point. What need is there for embryonic molts since the size of the embryo is a given, and the latter is placed within a rigid shell? And what need is there to bring the production of the cuticle forward in such a conspicuous fashion, and to engage the embryo in its renewal, with a significant metabolic cost, even should the animal be able to reutilize a good portion of the materials that the thin discarded film is made of? To state that all of this is undertaken on behalf of the greater safety of the newborn is to expect a little bit too much of the embryo's foresightedness. It would in fact constitute an undue, burdensome concession to that creeping finalism that is still given too much space in modern developmental biology. However, if the current "explanation" is not acceptable, then we need to come up with a better alternative. A couple of observations can help us in this regard.

The first concerns the nematode *Caenorhabditis elegans*. Its cuticle is formed precociously, as in arthropods. But, among the numerous mutations researchers have listed in their *C. elegans* cultures, there is one that results in the cuticle not being produced. This is therefore an excellent starting point for attempting to understand if it already has a role in its first appearance as embryonic cuticle. And in actuality it does. In mutants deprived of the cuticle, development is arrested prematurely, but for a reason that has nothing to do with the two best-known functions of a nematode's or arthropod's cuticle: a protective function against external agents and a role as internal skeleton on which the muscles can find a point to attach themselves. Much more simply, those embryos of *C. elegans* deprived of a cuticle assume irregular shapes, almost as if they were missing a mold, or, better, a sufficiently rigid container that would allow the hundreds of cells that constitute the embryo to assume their correct position. Without the cuticle they tend to drift. What probably occurs is that some of the interactions between neighboring cells, which in normal embryos guarantee that each cell assumes a functional identity that works harmoniously in the organism's overall economy, do not take place. In other words the embryonic cuticle (in *C. elegans* at least) is necessary for the embryo to assume, or maintain, the correct shape. Consequently the presence of this cuticle seems to be necessary from its inception and not, finalistically, with an eye to the situation (a free life, outside the shell) that the animal will only have to confront at a later stage.

We should then make a second observation. This time our inspiration will be derived from insects: from *Drosophila*, for example. During its entire life as a larva, the fruit fly does not undergo many cellular divisions. Those easiest to document are the ones that affect its epidermis, the thin layer of cells that envelops the body and that, on the outer surface, forms the cuticle.

Usually, counting the epidermal cells of an arthropod is an easy task because very often (in certain parts of the body at least) the outlines of each cell are stamped on the cuticle. If viewed when significantly enlarged, the cuticle looks like a floor covered with polygonal tiles, the so-called scutes. In the interval between one molt and the next, the number, size, and shape of these scutes do not vary. But what happens during the molts? On average, the size of the scutes (and therefore, we can suppose, that of the epidermal cells) does not change. Their number, instead, increases. This means that a certain number of epithelial cells have undergone division. It is not difficult to estimate how many cells have divided, since the average size of the epithelial cells is constant, which is the same as saying that for flooring one always uses tiles of the same size.

Let us suppose that before the molt one part of the body (the third segment of the thorax for instance) was 1 mm in length and that after the molt this same body part measures 1.1 mm, while preserving its shape. We can then estimate that after the molt the epidermis is formed by a number of cells that is $(1.1)^2 = 1.21$ greater than before. Therefore 21 percent of the cells that previously formed the epidermis must have divided in the meantime.

But these divisions are restricted to a brief time interval that just precedes the discarding of the old cuticle by the larva that is about to molt. Thus, another possible function of the cuticle becomes apparent, a function it would perform regardless of the form or the development phase the animal is in, and regardless of the presence of muscles that might be attached to it. This function would consist in more or less efficiently counteracting cellular division. In the overall economy of an animal enveloped in cuticle, this would enable a stabilizing of form in addition to (often, but not always, as we have seen) limiting an increase in size.

As in the case of the role of the cuticle in stabilizing the shape of the body that we observed in *C. elegans*, this ability of the cuticle to counteract cellular division is a property that can already have its importance during the life of the embryo. It is indeed possible to advance the hypothesis that the production of cuticle by the cells that envelop the body came about (and, in a certain sense, the mechanism that is still producing it in the course of the embryonic development of each arthropod and each nematode) as the result of a sort of competition between the different cell lines within a single embryo. To better explain this concept, however, it is necessary to digress.

Cilia and Mitosis

Precisely because they are equipped with a cuticle that is periodically renewed, nematodes, arthropods, and some other minor zoological groups occupy a special position in the animal kingdom. The presence of a cuticle is, in fact, hardly reconcilable with the presence of cellular cilia. These cilia are those diminutive structures, typical of aquatic organisms, that allow a paramecium (a ciliated protozoon), for example, to move in the water, or a sedentary animal like an oyster to create a water current capable of moving the suspended organic material that it feeds on toward its mouth.

Because of our nature as terrestrial animals we tend to too easily undervalue the diffusion and importance of cilia, but we should remember that they are present, and play important roles on the epithelia that coat our respiratory and genital tracts. The importance of ciliated cells is much greater in many marine and freshwater animals. The sea is quite particularly full of animal species that pass a juvenile stage of development as diminutive ciliated larvae. These are very frequently larvae with exotic shapes, and it is impossible to guess what adult forms these correspond to unless one has the opportunity to follow their metamorphosis. Even given this diversity of shapes, these larvae have two common traits: the diminutive size and the presence of cilia that coat a more or less extended part of their bodies. These cilia are extremely useful in moving food toward the mouth but—generally—less useful for locomotion, since the diminutive size of these larvae generally leaves them at the mercy of waves and currents. But even the cilia of these animals, just like the cuticle of arthropods and nematodes, appear long before these larvae begin their independent life. The problem is therefore to understand why the cilia are present in an embryo that does not move and cannot create a water current to carry organic particles suspended in the water toward it, from which it might derive nourishment.

The traditional explanations generally, and in this case, refer to functional cilia, which, if present, can be usefully employed in a subsequent stage of development. As an alternative, with a gaze directed at the geological past rather than at the future individual existence of what is now an embryo equipped with cilia, zoologists have attempted to explain the existence of ciliated embryos by proposing that they preserve the memory (today, perhaps, no longer functional) of the ciliated shape already present in a postembryonic free-living phase of some remote ancestor. An atavistic trait, therefore, that was not erased and that, perhaps, should be conserved precisely because—as we said regarding the arthropods' cuticle—if the characteristic is already present in the embryo, it will also be present in the newborn animal. But in this case, as in the case of cuticles, these traditional "explanations" don't really explain much. They appear rather to be ad hoc hypotheses whose degree of reliability it is impossible to verify. Popper would have called them not quite scientific.

A better explanation is however possible, and in the case of the cilia of embryos it was proposed by Leo Buss in 1987 in a juicy booklet entitled *The Evolution of Individuality*. Let us follow his reasoning, at least in outline.

We can start with the fertilized egg, which is subdivided into a growing number of cells, those that embryologists call blastomeres. First two, then four, eight, and so on. As long as there are only a few blastomeres, they face the embryo's surface, which one can imagine as spherical (as it in fact is in a great many animals). This generally means that all the blastomeres can maintain direct contact with the outside world because the eggs, in the majority of marine animals, do not have a shell, and the embryo is at best protected by matter with a fluid, gelatinous consistency. Eggs with shells are generally a prerogative of terrestrial animals.

As the divisions proceed, the number of blastomeres increases accordingly. The size of each individual blastomere becomes progressively smaller, because the overall mass of the embryo, which does not obtain nourishment from the outside, remains unchanged. Consequently the moment soon arrives when it is no longer possible to keep all the cells in a single layer, each with a little "face" exposed to the external world. Instead an ever increasing number of cells, which form the embryo's internal mass, find themselves covered by a certain number of cells that remain on the surface. Therefore, sooner or later, a conflict arises between the outer and the inner cell populations. It is certainly true that they both derive from a common ancestor (the fertilized egg) and that they therefore share common genetic information. But they live in different micro-environments, they receive and send different messages, and they express different genes and will gradually assume different specializations. The rhythms at which they divide are also different. And it is precisely here that the cilia, which only the external cells possess, once more come into play.

These external ciliated cells, in fact, are incapable of subdivision. Or, to put it more precisely, they are not capable of subdivision so long as they are equipped with cilia. Once the cilia are gone, they can once again subdivide. This is most probably due to the fact that the centriole, the cellular organelle that is at the base of each cilium, is the same structure that is the basis, in a cell that is starting to subdivide, for the formation of the mitotic spindle, that bundle of thin proteinaceous tubules along which the "halved" chromosomes destined to be the nucleus of each of the descendant cells will migrate, in the direction of the opposing poles of the spindle itself. It is therefore reasonable to think that the centriole cannot contemporaneously take charge of both entities: in other words the cilium and the mitotic spindle.

One thing at a time. Therefore being outfitted with cilia and undergoing cellular division belong to different moments of the life of a cell.

Returning to our embryos, where the cells of the most external layer are coated with cilia that do not seem to have any useful function, we should ask if some entity might not be interested in preventing these cells from dividing too often. Our suspicions, naturally, point toward the interior cells, whose access to the external world is preempted by the exterior cells. From their point of view, if I can express myself in this fashion, there are already more than enough exterior cells, and it is not in their interest that these continue to multiply freely. It is better to apply the brakes. In order to do this it may be sufficient to convince them to produce cilia.

Cilia, therefore, already have a function in the embryo's economy, a function that has nothing to do with those functions (above all related to locomotion) that are usually associated with cilia and can be put to use (even if *not* for the benefit of the ciliated cells themselves) from the very first moment that the cilia make their appearance on the embryo's most external cells.

Cilia and cuticles. Two alternative stories because usually a cell enveloped in cuticle does not manage to produce cilia capable of moving in water. Two stories that are perhaps equivalent, and equally significant. Stories of animal characteristics that appear, in the course of an animal's development, in too precocious a phase to be useful in the manner in which a cuticle or cilia are useful to an animal that has completed embryonic development, and that, in order to live, must move, feed itself and defend itself from enemies. In both cases the traditional explanations are drawn from the bottomless pit of finalism that developmental biology continues to employ. But it is time to take development seriously, to use the subtitle of a recent book by Jason Scott Robert.

Taking Development Seriously

Taking development seriously means to finally stop viewing it as the process that prepares the animal or plant for its adult existence as if the other stages, those that do not coincide with adulthood, could not or should not be interpreted according to the specificity of their condition, whatever its duration. One probably tends to privilege adulthood because it is the stage during which (normally) reproduction occurs. Without an adult who reproduces, all the previous history of the individual, from the egg on, would be destined to disappear without a trace. Not to mention the significance that the adult, with its rich cache of lived experiences, comes to assume in that rather special animal that is the human species. So, on one goes with an adultocentric perspective on living things.

It is not difficult to understand that, within such a perspective, it is not even possible to formulate the kinds of questions we posed in the last several pages, about the cuticles and the embryonic molts of arthropods and nematodes, or the cilia that coat the outer cells of an embryo. We would have limited ourselves to asking, as always (with a few laudable exceptions, like Leo Buss), what these embryonic cilia and cuticles could tell us about the cuticles and cilia of remote ancestors whose postembryonic stages, so one would claim, are recapitulated in the course of the embryonic development of the current forms. Instead we asked ourselves what purposes these cilia and cuticles could serve in the embryonic stage in which they appeared, and we saw that this question had a plausible answer: certainly not conclusive, as it requires comparisons and verifications. Maybe this is only the starting point for an entire research program, but one that already has some creditable claims. It confronts the developmental phenomena for what they are, *hic et nunc*, without unwarranted projections toward a future (the adult condition) that one claims is programmed, and without appealing to the nebulousness of an evolutionary history whose detailed reconstruction we do not in actual fact possess, though we claim to.

Ontogeny and Phylogeny

This leads to the difficult discussion about the relations between ontogeny and phylogeny. Ontogeny is the development of the individual from the egg on; phylogeny is the history of the species to which the individual belongs, a history that for simplicity's sake (perhaps exaggeratedly so) is often reduced to an ideal gallery of ancestors, starting from the most remote.

In these pages it is unnecessary to review the history of the theories that have been proposed about the relations between ontogeny and phylogeny, or to expound on the objections and reservations that have been advanced in this regard, or to discuss the new formulations that from time to time have been suggested by the advances in our knowledge as regards both ontogeny and phylogeny. In 1977 Stephen Jay Gould devoted what I think is his best book, *Ontogeny and Phylogeny*, to these problems. For readers interested in an accurate historical analysis and a good introduction to the modern debates on the topic I suggest they consult this book. In any case, this is an issue that cannot be easily avoided in a book devoted to evolutionary developmental biology.

The best known formulation of the relationship between ontogeny and phylogeny we owe to the German zoologist Ernst Haeckel; it has become famous with the name of "fundamental biogenetic law," and it can be concisely summarized in the following terms: ontogeny recapitulates phylogeny. It is rather like broadcasting a soap opera. At the beginning of the second episode one briefly summarizes the first. At the beginning of the nth episode, there is a brief summary of the preceding episodes, devoting slightly more air time to the last episode aired. In nonmetaphorical terms, Haeckel's biogenetic law presupposes that during the course of evolution the sequence of developmental stages an animal passes through is transmitted, basically unaltered, to its nearest descendants, with the possible exception of some final addition. In other words in the course of its individual development the descendant undergoes all the stages of development the ancestor has undergone, except for the final addition. Thus, what had been the anatomical organization of the ancestor as an adult becomes, in the descendant, an immature pre-adult stage, that evolves into a new stage (adult), for which there is no equivalent in the ancestor. The frog's tadpole, for instance, would embody the reminiscence of an adult ancestor of this frog, an ancestor that lived in water and breathed by means of gills (in other words, a fish). With the birth of the first amphibian, this aquatic phase would have become "compressed" and reduced to a juvenile (larval) phase, allowing room for the final addition of a new adult, with lungs and four legs.

Certainly if one could trust the validity of Haeckel's biogenetic law, it would be easy to reconstruct evolutionary history, even in the total absence of fossils. And this would be a fortunate turn of events, since only the animal's rigid parts (skeletons, teeth, and shells) easily become fossilized, and many animals are in fact made entirely of soft tissues that leave fossil traces only in exceptional circumstances. Unfortunately, however, we cannot entirely trust the biogenetic law, for a variety of reasons.

One of the limits of this "law," which Haeckel himself was well aware of, resides in the fact that evolutionary novelties are not all contained in the final part of an individual's history. Not everything can be reduced to a series of final additions. Consider, for example, the so-called egg's tooth, a rigid formation atop the chick's beak, which allows it to crack the egg shell, thus opening the road to the wide wide world. It is improbable that this tooth (which falls off shortly after birth) was present in some ancestor of our chicken and that in the course of evolution it was transformed into a temporary structure that the animal loses before reaching its new adult condition. It is instead more reasonable to think that it represents an innovation that evolved simultaneously with the production of eggs with a rigid shell. Beginning with its origin, therefore, the presence of this tooth is limited to a brief existence around the time of hatching, in other words to the moment during development in which this structure can actually be useful.

Once we have recognized the existence of evolutionary novelties that differ from the final addition of new stages, however, it is difficult to think that the exceptions are limited to the simple insertion, within the succession of developmental stages, of pointlike and easily recognizable innovations like the chick's egg tooth. No one can keep us from thinking that the sequence of developmental stages the ancestor undergoes could be modified in different ways, for instance by entirely erasing certain stages, or by upsetting their order.

We will be able to discuss this possibility of modifying the sequence of ontogenetic events and its role in determining the appearance of important evolutionary innovations again further on. For the moment we should be content to clarify that Haeckel's principle (ontogeny recapitulates phylogeny) was based on an excessively simplistic and reductive model of the manner in which an animal's development can be modified during the course of evolution. But there is another and more radical criticism that should be directed at the so-called biogenetic law. It presupposes a view of development that can be defined as adultocentric. And it is ironic to think that Haeckel had closely examined some facts that could have led him to a quite different interpretation of evolutionary history. In fact Haeckel interpreted the chick's egg tooth in a manner that is analogous to the explanation that Leo Buss gave to account for the presence of cilia on the embryo's surface, or the one I suggested for embryonic cuticles and molts. Ernst Haeckel in fact highlighted these innovations (or cenogenetic characteristics, as he preferred to call them) only to underscore their exceptional character, and therefore the danger they constituted should one attempt to reconstruct phylogeny on the basis of ontogeny.

Chapter 7

Paradigm Shifts

Science and Language

Thomas Kuhn has made the search for "paradigm shifts" in the history of scientific thought less unusual, shifts that from time to time mark modifications in our attention toward new ways of reading the world, while the interest for the old problems only survives in the most backward fringes of the scientific community. These are shifts that affect even the language used by the research community to debate its findings and interpretations, and in nontrivial ways. The latter, linguistic aspect of scientific evolution should certainly not be underestimated. The greatest difficulties it causes are not due to the continuous introduction of new terms, created specifically as a vehicle for the new concepts, but rather to the new meanings that accrue in many old terms through usage.

Even the protagonists of scientific research themselves often have difficulty becoming aware of this continuous change in the meaning of words insofar as they limit themselves to everyday usage related to the specific problems and objects to which they devote their efforts. An awareness of the multiplicity of meanings that one term can assume (sometimes even simultaneously, within different scientific communities, between whom exchanges are often almost totally absent) frequently emerges at the moment in which scientists take a break from active research in order to become engaged in the writing of a textbook or, better still, to reconstruct a segment of the history of their own discipline.

A deeper analysis of these semantic shifts, however, is officially the prerogative of those who are by profession either historians or philosophers of science. It is sufficient, though, to be somewhat familiar with the biological literature of the last decades to capture the repeated and frequently overlooked changes in meaning that have affected some of the key terms used in this book.

Gene, Genes

The most important example is the use of the word *gene*. From a certain point of view, one can easily understand that the progress in the understanding of the structure and functions of hereditary materials present in each cell has entailed repeated shifts in the meaning of this term. Forty years ago, for example, in the immediate aftermath of the deciphering of the genetic code, it seemed one could unambiguously identify the gene with that succession of nucleotides (the iterative units that make up a molecule of DNA or RNA) that corresponds to the succession of amino acids constituting a specific protein. One used to say: one gene, one protein. But molecular biology soon demonstrated that DNA, in addition to the segments that exactly correspond to the proteins (the so-called encoding segments), also contains other segments with a regulative function pertaining to the transcription of genetic information: in other words to the "reading" of a gene, which entails the construction of a complementary molecule of messenger-RNA (mRNA). A new problem thus arose: How should these regulative segments be treated? Are they part of the same gene to which the segment that codifies a protein belongs, or are they, in some sense, "accessory elements," to be considered separately from the genes themselves?

Semantic problems of this nature in effect revealed (even though many protagonists of this paradigm shift only belatedly became aware of these changes) a profound change in the direction of the study of heredity. Before modern molecular biology developed techniques capable of identifying the structures of the molecules to which the cells' hereditary information is tied, genetics had remained, for all intents and purposes, a science of the transmission of traits. This discipline's problem sets were concerned above all with the correlations between the traits of the parents and those of the offspring.

Crossing was its most important technique. Its idea of a gene was that of a "black box" that passes, usually unchanged, from one generation to the next (with its indefinable capacity to control the offspring's characteristics) according to the simple mechanisms revealed by Mendel, with the addition of a series of ancillary principles discovered in the course of the following half century of crosses, counts, and calculations.

Starting with the formulation of the double-helix model of DNA in 1953, genetics could, however, begin to look inside those that had, up to that point, remained simple black boxes. The genetic material would soon reveal both the relative simplicity and the basic universality of the "code" that ties it to the proteins synthesized according to the information it contains, and the enormous complexity of its organization. This organization, as mentioned, actually goes far beyond the simple succession of segments (the genes of the old genetics), each corresponding to a protein and therefore (by means of this succession) to a specific set of characteristics of the animal or plant in question.

This entry of the molecular structure of genetic material into the scientific world, with all the developments that have followed, including modern genetic engineering, was decisive in launching a chapter in biology that until that point had simply been outlined as a goal to be striven for someday, once the occasion arose. This new chapter is dedicated to the study of the manner in which the genes' expression is translated (by means of their transcription onto molecules of mRNA, followed by the synthesis of the corresponding protein molecules), during development, into the progressive differentiation of various cell kinds, and subsequently into the ordered appearance of the different organs: in a word, into the unfolding of biological forms.

From the old genetics of the transmission of characteristics, one passed, by way of a radical paradigm shift, to the genetics of development. From this perspective, the gene was no longer a simple black box, unchanged until a mutation somehow succeeded in modifying it. And it could no longer be the sequence of nucleotides corresponding to a specific protein, or better to an enzyme, as was taught forty years ago. A lesson meant to highlight the importance of this relationship (one gene → one messenger RNA → one protein) was actually defined as the central dogma of molecular biology in the course of its triumphant advance. One thus passed from the Mendelian idea of gene—as a not well defined factor that is passed unchanged from one generation to the next, and on which the presence of a specific characteristic in the individual carrying it depends—to a more recent idea, in which the gene, in addition to acquiring a precise molecular identity, comes to lose its property of indivisibility and exhibits a complex structure, suggesting many things about the manner in which it functions.

But this semantic shift is nothing more than the most visible manifestation of a more extensive network of problems: when confronting it one is tempted to use Luigi Pirandello's formulation, and say that the gene is one, none, and one hundred thousand. All of this seems sufficient to make us ask to what extent biologists educated in different research traditions such as population genetics, molecular biology, and developmental biology are aware that, when they talk about genes among themselves, they are really often talking about entities that differ very markedly from one another, of concepts, in other words, that are part of research paradigms without much in common. The necessity to put some order in this field really becomes acute when one confronts problems at the borders of traditional biological disciplines. And this is why evolutionary developmental biology can become a terrain of rigorous critical revision, also as regards the topic of the gene.

Busillis

However, in order to address other concepts, it is not necessary to ask Kuhn or some other philosopher of science for advice. Many concepts, and, to be more precise, many among the most popular ones and those one should seem able to define unambiguously, are also undergoing a crisis, without waiting for the relevant lines of research to undergo a traumatic paradigm shift. The problem, in this case, is that these concepts seem usefully applicable only within a narrow framework, while they are revealed as painfully inadequate when employed in a wider context.

What too often renders a systematic application of some concepts too problematic is that they derive from a prescientific classification of natural phenomena, a classification, in other words, that was developed to order a limited number of objects and processes, most frequently those concerning human beings or the animals that most resemble them (mammals, vertebrates). In this limited context, perhaps, they still have a certain validity. However, the further removed one is from their initial domain of application, the more the difficulties increase. The story of that cleric from bygone times comes to mind, who, on the last line of a page of a large illuminated missal, where the copyist had aligned the words without separating them clearly from one another, read a brief series of letters that seemed easy for him to understand: INDIE, in the day. Then he turned the page, his horizon expanded, and the new column of the text began with eight letters that our cleric had never faced before: BUSILLIS, he read. And he no longer knew how to proceed. Obviously his mistake had been to rely too blindly on the caesura represented by the page break. He had, without any precise justification, given it the value of a break not inferior to that which separates two words in the same sentence. And he made a mistake. One word—DIEBUS—actually straddled the two pages. The break he should have made was actually waiting for him three letters after the place where he had made it. And there were still many pages to be turned. You, young cleric, do not even know how many. You would be wiser if your manner of separating words from the old sacred text became freer and more courageous.

And at bottom this is the situation for concepts such as growth, development, reproduction, regeneration, larva, and individual. These are concepts that have ancient roots, and are increasingly ambiguous in their modern applications; and they slowly also affect less ancient concepts borrowed from everyday language and born inside the domain of modern biology. Among these are the concepts of tissue and segment. To this list of concepts in crisis, of ancient or more recent origin, we could add species and homology. To talk about species, however, would take us far from the topics of this book, and we shall return to homology in chapter 8.

Tissues

The wares on display on the butcher's counter can offer a first, rough introduction to the science of tissues. The appearance and the desirability of a slice of muscle, of a piece of liver or brain, of a bone fragment with the tendons attached, is different for almost all of us. Each one of these materials suggests different recipes to transform the freshly butchered product into a dish appropriate for modern human beings.

Histologists naturally do not stop at this first level of analysis. In the most apparently homogenous slice of red meat, they would be ready to point out, next to the prevalent muscle tissue, a whole series of structures composed of nerve tissue or endothelium (the blood vessels' wall) and even blood, a tissue made of cells detached from one another, because immersed in the liquid matrix that is plasma. An analogous, if not even more detailed analysis, could be offered for the other samples.

It was the systematic use of the microscope, combined with the application of ever more refined techniques that, with the use of special dyes, enabled us to highlight the differences that exist between the different cellular constituents of organisms' bodies, and allowed histology to develop. It was in this fashion that we discovered the differences between smooth and striated muscular fiber, between epithelial and connective tissue.

Many times, it seems easy to establish correspondences between the structures of two animals so distant from one another, that, on a macroscopic level, comparative anatomy has difficulty in recognizing the equivalence between the organs of the one and those of the other, but this is not always the case. There are relatively frequent occasions in which our classifications run into trouble, and as a way out we have to be satisfied with the creation of new hybrid classes, as in the case of the myoepithelial cells (partly muscular fiber, partly epidermal cell), or the neuroepithelial ones (partly epidermal cell, partly neuron) that we find in the body of the hydra and other cnidarians. And this is not to mention neurosecretory cells, which are widely diffused in the most diverse zoological groups; their discovery, by now in the relatively distant past, demonstrated the insufficiency of models where the nervous and endocrine systems seemed to represent two totally separated worlds.

Apart from the intrinsic limits of every attempt at classification for which, from time to time, it is best to reexamine the relevant domain of application, one should here observe that evolution does not only involve organs and apparatuses, but pervades all levels in the organization of living organisms, beginning precisely at the cellular level. In fact, many significant developments in the history of evolution were contingent precisely on these changes in the properties and functions of individual cells. Examples of this development are the capacity of some cells to remain attached to one another—a fundamental property in the passage from a unicellular condition to a multicellular one—or that of directional growth in response to certain stimuli that can be observed for instance in the hyphae of fungi, in the neurons of animals, and in the vegetative apexes of green plants. Consequently, we cannot expect that a classification of cellular types born to serve the needs of medicine and, in any case, founded on the characteristics of the cells of mammals, be applied also, without significant problems, to insects, mollusks, and sponges. Nor will it be sufficient to lengthen the list by including the cellular types that are remotest from those of mammals, for instance the stinging cells of jellyfish or the cells that in many sponges produce rigid calcareous or siliceous needles, with their characteristic shapes. We should instead expect cases like those of the myo- and neuroepithelial cells of the hydra, in which we find, in the same minimal unit of living matter, structural aspects and functional properties gathered together that in other animals are instead spread across distinct cellular types. To apply a classification of tissues that was originally intended for human beings and, perhaps, for other vertebrates, to an animal very different from us, often results in significant difficulties,

difficulties that can only apparently be resolved by means of the lexical acrobatics currently used to describe the organization of tissues in the little polyp.

More on Segments

Until recently, most zoologists continued to accept as natural the close relationship between annelids and arthropods that had led Georges Cuvier, at the beginnings of the nineteenth century, to include both among the Articulata, one of his four great subdivisions (*embranchements*) of the animal world. Undoubtedly one the most evident characteristics of these animals' organization is the division of their body into segments. We have already discussed this issue several times in this book, especially insofar as regards the singular penchant of scolopendras, and similar organisms, to possess a specific number of segments, and also when discussing the basically invariable number of segments that constitutes the body of leeches.

The most obvious difference between the segments of arthropods and those of annelids is that the segments in the latter are devoid of articulated appendages, while generally arthropods do exhibit them, at least on part of their body. Another difference between the two groups is the thick and sometimes rigid cuticle that usually protects the body of arthropods, while in annelids the cuticle is thin and flexible. A dissection, naturally, would reveal many other differences, but it would simultaneously show us that this segmented structure, which is externally visible both in annelids and arthropods, has an equivalent in the serial arrangement of some internal organs in both cases. This circumstance would therefore seem to indicate that the similarity between annelids and arthropods, insofar as they are animals whose bodies are subdivided into segments, is something to be taken into serious consideration. The matter, however, is not so simple.

On the one hand, annelids and arthropods are not the only animals in which the body's main axis is subdivided into iterative units. On the other hand, perhaps the segments in annelids and arthropods have a different origin. A body articulated into segments is probably something that has developed on several separate occasions in the animal world.

Let us consider our own backbone, for instance. It is impossible to deny its serial structure, which becomes even more evident in the case of many fish, in which the vertebrae are structurally much more homogeneous than in the human skeleton. Moreover, in fish, the muscular masses in the trunk and the tail (if you want to check, all you need to do is to put a sole or a hake on your plate) are subdivided into short units that repeat following the same spacing as the vertebrae and their long spinous processes. Therefore, from a certain point of view, vertebrates can also be considered segmented animals. This circumstance, however, does not seem sufficient to make us consider the vertebrates a group of animals that is particularly close to annelids or arthropods (an opinion that has only been defended by a negligible number of zoologists in the last century and a half).

And what should we say about the structure of a tapeworm, whose body is mostly constituted by a series of iterative units? These units certainly have a much greater autonomy than can be seen in the segments of an earthworm or a millipede. In fact in many species of tapeworm they can detach one by one, with the whole load of gametes they contain, without disturbing the functioning of the remaining part of the worm. One should underscore the fact that the tapeworm is devoid of a digestive tract and, therefore, of one of the structural elements that, in the earthworm or in the millipede, transform the series of segments into a unit, since the intestine passes through their entire series, from one extremity of the organism to the other.

For all these reasons, and because of the manner in which the tapeworm produces these serial elements in its body, zoologists have generally refused to designate these sections as segments, preferring instead to call them proglottids, a term that is not applied to the iterative sections of any other animal. In a certain sense, one seems to be able to read between the lines that tapeworms are not worthy of segments. If it is true that they derive from nonparasitic flatworms whose bodies don't present any trace of subdivision into segments, the architecture of their bodies must be described in terms that differ from those we use for annelids and arthropods. Tapeworms have nothing in common with these two groups of animals, in other words with Cuvier's Articulata.

Apart from terminology, it is difficult to imagine an evolutionary scenario that would trace tapeworms, annelids, and arthropods back to a common ancestor whose body was articulated into segments. In other words, proglottids are really an invention of tapeworms. At this point, however, we should take a step backward.

Origin, Form, and Function

A body axis formed by periodic units that succeed one another, from one end of the animal to the other, seems to have appeared in at least three different cases: in vertebrates, in tapeworms, and in Articulata. If this is the case, should we really wonder that a segmented organization of the body actually appeared many times? More specifically, are we really certain of the equivalence of the segments of annelids and arthropods? If the regular repetition of organs that can be observed along the principal body axis of these two zoological groups is the result of independent evolutionary innovations, the principal anatomical characteristic that sustains the notion of Articulata would disappear. Instead of being the closest of relatives, annelids and arthropods could in fact entertain different relations within the animal kingdom.

This objection has actually been raised repeatedly. As of this writing, moreover, the issue does not seem to have been definitively solved. There is a small number of zoologists (and some authoritative ones belong to this group) that has not lost its faith in the relationship between annelids and arthropods proposed by Cuvier. For a growing number of their colleagues, however, the segments of annelids are not the same thing as the segments of arthropods, and Articulata should therefore be considered an artificial group. In the long run it is probable that this will be the victorious view. More specifically one should take note of the fact that our current knowledge of developmental mechanisms seems to indicate that the segments are constructed differently in annelids and arthropods, and this seems a good argument in support of the thesis that the last ancestor the two groups had in common, probably an extremely remote one, did not have a segmented body.

This conclusion, interesting in itself within the context of an improved knowledge of animal evolution, doesn't exhaust our interest for segments. If they were actually reinvented many times, if—in other words—there is no unique recipe for building them, perhaps at this point we should ask ourselves what a segment actually is. This is a question that—like many others in biology—does not necessarily have one unequivocal answer.

In many cases, segments are structural and functional units that have a precise role in locomotion. In the earthworm, for example, segments are mostly cylinders of equal volume that can change shape (getting shorter and expanding, or getting longer and reducing their diameter) thanks to muscular contractions. And in scolopendras each segment of the trunk, which is loosely linked to the preceding and following segments, is equipped with a pair of legs whose movement is well coordinated with that of the other pairs.

In many other cases, however, segmental composition is only a morphological datum to which it is difficult to assign a specific functional value. For instance one commonly says that an insect's head is composed of the first six segments of its body, or that in leeches the posterior sucker is composed of the last seven. But any trace of segmentation is absent in the actual sucker, and its segmental composition is simply inferred on the basis of what is observable in the embryo, when the segments from the twenty-eighth on still appear somehow distinct. And in the case of insects, the segmental composition of the head, leaving the embryological clues aside for the moment, can be inferred from the presence of a series of appendages (the antennae, the mandibles, etc.), each pair of which is traditionally assigned its own segment. Therefore when we speak about the segmental composition of an insect's head or a leech's sucker, the segments we refer to are not distinct anatomical parts, but units of a structural composition that have been variously modified, integrated to one another, and sometimes perhaps eliminated. Very frequently only a trace of this or that segment that we imagine has become part of a specific body part is still actually present (for instance a nervous center, perhaps largely integrated with neighboring centers). Certainly, we say, these are modified segments, as can be expected in evolved, specialized animals: if we could only go back to the dawn of time, we would certainly find their ancestors with a body formed by identical segments.

But it certainly seems possible (or, rather, probable) that a similar type of modular animal formed by identical, complete, and distinct segments never existed, at least along the evolutionary line that leads to modern arthropods. When the division of the trunk into segments made its first appearance, it is very likely that the animal already had a separate head, and in this part of the body, it is unlikely that the serial repetition of organs ever affected the entire set of organs to be found in a typical segment of the trunk. The segment as a fundamental unit of the composition of the body in all these animals is, in all probability, nothing more than an abstraction of ours.

The undoubted advantage that many animals derive from their segmented bodily architecture, at the level of efficiency of locomotion was, probably, one of the reasons for the evolutionary success of the annelids, the arthropods, and the vertebrates. But the reasons for the first appearance of the division of the body into segments must be sought elsewhere. They must be sought for directly in the mechanics of development.

Parallel Worksites

To get some suggestions let us return to the leech. In the third chapter we opened a window onto the incredible developmental mechanisms that ensure this annelid will construct a precise, invariable number of segments. The reader may recall the activity of the teloblasts, that small group of cells from whose proliferation, in a brief period of time, the leech obtains the formation of thirty-two transversal cell striplets, as many as will constitute the animal's definitive segments. The production of these thirty-two segment outlines occurs in a serial manner, beginning at the anterior extremity. As soon as the first trace of the segmental architecture of the leech has been outlined, each of the thirty-two groups of cells begins to proliferate and differentiate, giving rise to sensory cells, nerve ganglia, muscular bundles, excretory organs, and so on. It is as if, within the little developing animal, thirty-two worksites are operating simultaneously, each capable of realizing all the structures of the little district that it oversees. Special local conditions may, at some point, limit, expand, or in some way modify the activity of some worksites. In some segments, for example, excretory organs will never be formed; in others, reproductive organs will be added. Altogether, however, the products of the activities of the various worksites will be more or less the same. It is possible that the "cellular worksites" nearest the anterior extremity of the animal's body may complete their work a little sooner than those following them, but in any case the simultaneous activity of so many proliferation and differentiation centers makes the construction of the entire animal particularly efficient and secure.

That the finished product preserves traces of the manner in which it was constructed, appearing in other words as a segmented animal, is another story. An important one, naturally, because this is where natural selection can enter the picture, helping the consolidation and the perfecting of a modular body architecture that can guarantee the animal's good performance during locomotion. But the first part of the story, the one that concerns the origin of the segments is—like many other stories we have concerned ourselves with—a story that above all concerns developmental mechanisms.

From this perspective, segments are no longer the more or less modified fundamental units of the bodily structure, reduced or fused together under pressure from natural selection. They are instead the trace (sometimes well preserved, sometimes not) of a developmental process that has proven itself particularly efficient and trustworthy precisely because it is distributed along a number of centers precociously organized along the embryo's longitudinal axis.

At this point it is perhaps best to return to our idea of worksites operating simultaneously, to ask ourselves whether it is really true that optimum efficiency is reached when the different worksites share all their activities equally among themselves, each one assuming responsibility for the area that is under its control. If, for example, we want to build a highway, it certainly behooves us to open a worksite for each bridge or viaduct we want to build, and this worksite will therefore be responsible for a small segment and a very exacting operation; but for other simpler operations, such as asphalting, it would probably behoove us to open a lesser number of operative units, each one empowered to work on a longer stretch. For yet other operations, such as traffic signs, only one team traveling the whole length of the highway might be sufficient. The nature of the operations themselves (in other words the functional organization of the construction operations of the highway itself), or that of their products (in other words the functional integration of the different products that are its result), will, each time, suggest whether it is expedient to tie the different operations together or not.

Analogously when deciding if and to what extent the different sensory, locomotive, and excretory organs that repeat along the principal body axis should be coordinated, reasons tied to the logic of development, or to the functional advantages possibly associated with the mutual integration of the discrete repeating structures may prevail. This means that the segment is not an a priori structural block that can only be variously "sculpted" by natural selection. It is rather a point of arrival that is reached if and when developmental and adaptive logic favor the serial repetition of a structural complex, which is more or less varied and definable only once it is completed.

What Do We Start From?

Evolutionary developmental biology has therefore led us to a view of nature in which many traditional concepts of morphology—including those of tissue and segment, which seemed easier to define—appear in a new light. These concepts in fact no longer correspond to uniquely defined units whose evolution we can study or whose generative processes we can highlight. A plurality of structures, each one the result of intertwined developmental processes, now corresponds to each of these concepts. In the course of evolution these units have taken different forms that more or less correspond, often only imperfectly, to those elementary, almost archetypical, forms that we were used to tracing them back to. However this revision, which reveals the limits of the manner in which animal forms were traditionally described, does not stop at a critique of the concepts related to these structures, but can also be applied to the processes that produce them.

More specifically the periodization of development, in other words its subdivision into a series of temporal segments separated by important and uniquely recognizable events, becomes problematic. This could also seem to be a merely academic exercise, a playing with words, but this is not the case. Finding a reasonable way to subdivide the development of an animal into a precise series of stages is a necessary passage in view of an important operation: the comparison of the developmental histories of different animals. But it is not easy to achieve.

There are some events that seem to represent good points of reference: birth for example, or the achievement of sexual maturity. But here too there are problems. With birds, what we call birth corresponds to hatching, but what about with mammals? Is it legitimate to consider the entire intrauterine phase of the mammals' development and the sequence of events that separates the fertilization of a chicken egg from the moment in which the chick walks through the remains of the shell it has just broken, as equivalent? And what about the kangaroo and other marsupials, in which the offspring only remains in the mother's womb for a very brief period of time, reaching a stage similar to that of other new-born mammals only after its sojourn in the marsupium? We could try to get around the problem by designating the stage when the kangaroo is born as larval, but the problem of comparison remains.

We encounter similar difficulties with arthropods. In many species, once the egg has hatched, the body's appendages are completely developed and functional, and the little animal is ready to embark on an active life. In many other cases, however, the little arthropod that appears after hatching has an embryonic appearance, its appendages are short and not well articulated, and no movement is possible.

Another molt will be necessary before the little spider, or the little grasshopper, will be able to start their independent existence. What therefore constitutes birth in these little animals? Shall we decide to have it coincide with the moment in which the egg hatches, or with the beginning of an active life, disregarding the issue of whether the last molt preceding this phase occurred within or outside the egg?

It is clear that both choices are arbitrary: they only serve to limit the meaning of a term, purely for our convenience. But what ought to interest us, if we want to understand some biology rather than dither with lexical issues, is precisely the fact that different phenomena, such as the hatching from the egg and the attainment (by means of a decisive molt) of articulated appendages are not part of a unitary "package." Each phenomenon has its own mechanisms, its own developmental schedule, and, let us not forget, its own value for survival. The logics of development and of adaptation intersect each other in all phases of development, and in each phase they can favor either the coupling or the uncoupling of different phenomena, as we just observed.

Which Adult?

If birth is a moment in development that it is not always possible to specify without ambiguity, the same is true for the attainment of adulthood. Once more, arthropods can offer us good examples for discussion. In the case of a butterfly, a fly, or a scarab, the attainment of sexual maturity is accompanied by the transformation of the animal into what will become its definitive structure. After having undergone some larval stages and the revolutionary (but immobile) pupal stage, with a last molt whose results it will have to accept until the end, it turns into a winged insect. No further molts, and it will therefore also remain the prisoner of an external skeleton whose shape is not modifiable, and will no longer undergo significant morphological changes. In these arthropods, the passage to an adult phase of development does not seem to allow for any ambiguity. But we shouldn't be too certain about this. In some insects, the gonads are full of mature gametes before the last molt; in others, real maturity will only be reached at a later time, after the winged insect has been able to procure nourishment for itself. Can we therefore state that in both cases the molting from pupa to winged insect has exactly the same significance, from the point of view of attaining an adult condition?

One must further add that there are insects that do not stop molting once they have acquired wings. In fact in mayflies, there are two winged stages, and it is normally only during the last of these that the insect reproduces. In crabs, crayfish, and other arthropods, after first reaching sexual maturity, a number of successive molts actually take place. In the males of some millipedes, a first adult stage can be followed, after a molt, by a stage in which the animal is incapable of reproducing, unless this is followed in its turn by another molt and a stage in which reproduction is once again possible.

And if this set of examples is not sufficient in illustrating how imprecise the definition of the adult condition of an arthropod is, perhaps the mention of a peculiar kind of gnat will be useful: sometimes, instead of completing their normal metamorphosis and reproducing after having reached the winged stage, these gnats allow their eggs to reach maturity, giving rise to new individuals by means of parthenogenesis, in other words without fertilization, while they are still larvae—or, to be more specific, when their overall body architecture is that of a normal larva, while their gonads, instead of still being rudimentary, as usually occurs in larvae, are already in a state that can be defined as mature. I would be tempted to say that in this case the little creature is composed of body parts of different ages: the gonad of an adult, as previously stated, in the body of a larva. In gnats, in any case, reproduction does not always occur at the larval stage. Under certain environmental conditions, the larvae, instead of being child mothers, transform into pupae and then into "conventional" adults who are naturally capable of reproduction.

But the road along which gnats are headed seems to have led to even more disconcerting consequences in another group of insects. Paedogenesis, or reproduction by an animal at the larval stage, is also known to occur in a small American beetle, *Micromalthus debilis* (fig. 9). In this species as well, the fate of the larvae is variable, a func-

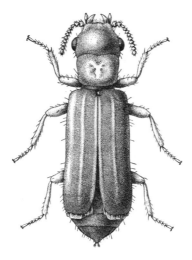

Figure 9. *Micromalthus debilis.*

tion of factors that are still not very well known. Some larvae lay eggs (and are then eaten by the offspring larvae), others instead continue their development until they become winged insects. One is tempted to refer to the latter as adulthood, but in this case we must use the terminology with extreme caution. It seems in fact that this little insect with the typical structure of an adult beetle (four wings, the first pair transformed into a couple of elytra) doesn't reproduce at all. If, as it seems, this is how things stand, then we are faced with an insect that, in what can on the whole be described as a larval stage, takes on the reproductive functions of an adult, whereas what on the whole, structurally, seems to correspond to the adult stage of other beetles, cannot reach sexual maturity. Who, then, is the adult, in *Micromalthus*?

Siamese Twins and Corals

We could continue for a long time to deconstruct many important concepts in biology whose usage seems more fraught with problems the farther we get from the organisms, or the situations, that these concepts were first introduced to refer to: organisms and situations that often have a lot to do with human beings or in any case with mammals, or, more generally, with vertebrates—an extremely important zoological group, and a very diverse one, which is, however, not faced with many of the functional and structural issues of other animals.

As long as we deal with vertebrates, for instance, we will not often have problems with the notion of the individual. The only occasions in which doubts might arise as to the possibility of precisely defining the boundaries between two individuals are those in which we notice the many "coincidences" that the life of two monovular twins is laced with. These doubts, however, become considerably more serious when fate denies the monovular twins complete anatomical separation—when, in other words, a pair of Siamese twins is born. Only the anguish for their (or its?) survival under these unusual circumstances can relegate to the background concerns about the semantic applicability of our usual notion of individual, with all the implications, including ethical and legal ones, that it entails.

But the borderline situation of Siamese twins, which it is all too easy for us to marginalize because of its extreme rarity, is the normal situation in which many animals find themselves, as is the case for the overwhelming majority of known species of coral. The tiny polyps, shaped like white stars, that adorn the branches of red coral are in fact identical copies of one another on the genetic level, and their relative anatomical and functional autonomy is never complete, since they are connected by common tissue. The problem of individuality is often avoided by zoologists, who give the entire group the name *colony*, whereas they prefer to give the name *polyp*, or even the more neutral name *zooid*, to the individual floret-animals, while avoiding the more compromising (or, in this case, less appropriate) one of *individual* instead.

A simple question of semantics, one could object, but one would be in error. The problems we encounter when trying to systematically apply the notion of individual are, once more, problems that have their roots in an intricate web of developmental mechanisms and adaptive opportunities. The biological individual, in the sense this term can be applied to a human being or a dog, is not a universally valid category, applicable with certainty to all our analyses of the world of the living. In this case also, as in the case of the articulation of the body into segments, or of the condition of adulthood, what we would normally consider a category applicable to the entire animal kingdom (not to speak of other living organisms) is only one of several possibilities that can be selected as the most suitable from the varied products that different developmental mechanisms are able to provide. In other words the biological individual exists when its realization is possible and its preservation advantageous: but this is far from being a universal necessity.

Spare Parts

Having its foundations shaken, the entire conceptual edifice based on traditional notions of reproduction, growth, development, and regeneration is showing other cracks that endanger its stability. We would therefore do well to intervene.

Many animals are able to reconstruct a part they have lost. This regenerative power is sometimes visible, as in the case of the lizard, which is capable of regenerating a tail if it loses its own, or in the even more glaring case of many starfish, where a single arm is capable of regenerating the entire animal. In our species, regenerative capacities are very modest, since they do not extend much beyond the healing of a wound. Insofar as regenerating lost parts is concerned, we cannot even regenerate the smallest phalanx on our fingers. At the opposite extreme we can place—more remarkable even than starfish, though their feats are certainly worthy of note—that little freshwater animal known as hydra. Hydra, like the mythical monster from the swamps near Lerna that had nine heads ready to grow back in case they were cut off. The zoologists' small hydra is in actual fact also a champion at regeneration. From a small piece of the animal, maybe a few hundred cells, the entire hydra can be regenerated.

The hydra is a small polyp, like the coral's "little florets," and though it is not a colonial animal, it is often difficult to establish where the boundaries of the hydra individual really are. In fact this animal usually reproduces by means of a mechanism known as gemmation. No eggs, no spermatozoa. Rather, along the side of the polyp's long cylindrical body an excrescence one day arises that slowly becomes longer and takes on the shape of a polyp similar to the original. After several days this "child" polyp detaches and begins an autonomous life. From that moment on we could fairly obviously say that there are two individuals. But at what instant, if such an instant exists, can we calmly say that it (they) passed from a single individual into two separate ones? Is it perhaps the moment in which the last relationship of anatomical continuity between what now appears to be two separate polyps is severed? Or is it possible instead that a large measure of physiological independence between the parent polyp and the one it produced by gemmation had been present for some time?

It is obviously improbable that these questions should have a univocal and sensible answer. The situation we discussed in the case of the coral seems to confront us once again, where any use of the notion of biological individual cannot but distort, at least to a certain degree, the current usage of the term. In the case of the hydra, however, there are some quite different issues to discuss, ones that go beyond the by now more usual uncertainties about the notion of individual. First of all there is the fact that the cells that constitute the hydra are subject to continuous, rapid replacement. A certain number of interstitial cells are distributed all over the body of the hydra, like a sort of stem cell that doesn't remain quiescent, waiting to be reawakened only in certain special cases, but always proliferating, providing cells that will be incorporated into different parts of the body, where they will differentiate each time in accordance with local conditions.

This continuous production of new cells compensates for a just as continuous loss of cells, which is particularly evident at the extremities of the long tentacles that surround the mouth. The net result of these opposing processes (the loss of differentiated cells and the birth of new ones starting from undifferentiated interstitial cells) is the maintenance of a constant shape over time, notwithstanding the animal's continuously changing composition. Like the river that maintains its shape even if the water that flows within its banks is never the same at two different points in time. Like Theseus' ship, which docks at many ports and has one of its damaged parts replaced at each one, until, after many stops, not a single plank or sail from the original ship remains. The paradoxes of continuity in change, the hydra teaches us, are not only food for philosophers.

If we think about the issue carefully, a lizard with a new tail or a starfish that regenerates from a single arm would be sufficient to make us hesitate. Are these, 100 percent, the same individuals that existed originally, before the trauma and the regeneration of the lost parts? Yes, perhaps, but we are more likely to agree with this in the case of the lizard, which ultimately only replaced its posterior appendage, than in the case of the starfish, where the surviving part of the original individual is actually smaller, in quantitative terms, than the regenerated one.

And finally, regeneration has many traits in common with forms of asexual, vegetative reproduction. The hydra's gemmation also belongs to this heterogeneous category of phenomena, but other animals can provide even better examples for a comparison with regeneration. There are the catenulids, for example. A nice name for a group of diminutive freshwater worms that mostly appear as little chains of developing animals: in the older ones the mouth is already complete, and they are then ready to leave the group they were formed from, while in the younger ones the outline of the extremities is barely traced, showing two slight narrowings of the little chain, one in front and one behind. We could even aid in this process of division, cutting the chain at any arbitrary location, and the result would not be much different from what nature accomplishes on its own with these animals.

Detachment and proliferation, proliferation and detachment. There do not seem to be too many differences between regeneration and asexual reproduction, except in the temporal sequencing of processes and in their greater or lesser regularity and foreseeability. Nature, of course, is a little more complex, in part because not all animals capable of asexual reproduction are also capable of regenerating, not to mention the fact, which is more easily under-stood, that among animals capable of regeneration only some also practice asexual reproduction. The fact remains, however—and this is the message that should be driven home, useful as it is for our reflections on development and evolution—that the boundary between these different phenomena is not as clear as we all too often would like to believe.

Chapter 8

Comparisons

Horns and Antlers

Modern evolutionary developmental biology has brought some figures back to light that had been relegated to obscurity by their more fortunate contemporaries and has reevaluated some ideas that had been put aside as mistaken, or at the very least unproductive, for a considerable time. The figure most worthy of note in this ideal gallery of reevaluated characters is certainly that of Étienne Geoffroy Saint-Hilaire. With renewed attention to the work of this great French scholar of the Napoleonic era, an interest in the set of daring anatomical comparisons that had been suggested to him by his deep conviction in the underlying unity of structural plan in all animals has also been reborn.

In times much closer to us, the possibility of recognizing what we can legitimately call expressions of variants of the same gene has opened horizons for comparison that go well beyond even the most daring attempts made by Geoffroy and his pupils. In writing this I do not mean to say that the expression of the relevant genes constitutes a causal explanation of the forms of these living organisms. As I mentioned in chapter 4 the expression of *Pax6* is not the *cause* of the eye's development, just as the expression of *tinman* is not the *cause* of the heart's development. Moreover, the comparative method in biology does not have the very ambitious goal of revealing the causal chain that leads to the development of the eye or the heart; but it can help us reconstruct the history of phylogenetic changes that gave rise to the various types of eyes and hearts we see in animals today.

But the first objective a comparative analysis should aim for is to understand what it is legitimate to compare and why. This problem, which appears to be very simple, actually hides many deceptions. What criteria, in fact, should we apply or privilege, when choosing the terms of comparison?

It might in fact appear sensible to compare our upper limbs with the wings of a bat or the front legs of a horse, even though these appendages differ quite considerably in their external shape and, even more visibly, in their function. Lengthy anatomical studies are not necessary in order to know that the skeletal scaffolding of these various limbs is very similar and that the manner in which they are tied to the skeleton of the trunk of the animal in question is also similar. It may similarly seem sensible to compare the delicate wing of a fly with the coriaceous elytron of a beetle, even though in the latter one cannot distinguish the innervations that support the membrane of the first. In this case also, the position of the appendage relative to the body, and the presence of a comparable articulation relative to the thorax (more precisely: relative to the second segment of the thorax) can certainly encourage us to continue.

These two examples alone are sufficient to show that two anatomical parts do not have to be similar either in terms of appearance or function in order to deserve comparison. Naturally nobody would object if there were *also* a strict resemblance between the two, but the rationality of the comparison does not depend on this. For example the front leg of a horse and the front leg of a zebra are very similar, to the extent that it would make no difference whether we take one or the other as a model for comparison with the wing of the bat or the upper limb of a human being. The fact is, rather, that the resemblance between the leg of the horse and that of the zebra is such that, practically speaking, the interest in a comparison between the two is reduced to zero.

Nevertheless, a resemblance, even a marked one, can be deceitful. What to say, for example, of the showy appendages that adorn the heads of many ruminants, most especially the males? It is not difficult—above all, for those who already carry the burden of half a century on their shoulders—to visualize those once fashionable hunting trophies that transmitted the memory, or the myth, of the adventures of some ancestor in Africa or, at least, in the wilder parts of the Alps from generation to generation. These types of collections were on view even in museums of natural history but thankfully have disappeared almost entirely from public display rooms, and have ended up in corridors and storage rooms—as is the case with Milan's Civic Museum of Natural History, where one of the most recent directors renamed one of these spaces the "hornidor."

In actual fact the problem here really is one of horns. The bull and the bison, the buffalo and the antelope all have horns, but we should pause a moment before adding the stag, the roe, the elk, and the reindeer to the list. As far as these animals' head ornaments are concerned, it is more appropriate to speak of antlers than horns. And we are not simply trying to be picky. A buffalo's horns and a stag's antlers don't have much in common except for being located on the head. It is sufficient to point out that the stag's antlers, unlike the buffalo's horns, do not have an inner core made of bone (an extension of the cranial bones) and that they are shed once a year. When such important differences emerge, using two names for the different appendages is the least one can do. They are obviously not organs of the same type, whatever this expression actually means. Naturally this has been known for a long time, since before Darwin and the birth of experimental embryology, genetics, and molecular biology, to say nothing of evolutionary developmental biology. But it is precisely with the birth of this latest discipline that the applications of the comparative method in biology have been given new life and are now directed at objectives that were previously thought to be, more or less, the exclusive prerogative of visionaries.

Homology

The key notion in this discussion is that of homology. The same homology that Belon, without using the term, saw between the bones of the human skeleton and those of the bird. It is the same homology that allows us to state that the wing of a sparrow and the fin of a dolphin are, from a certain point of view, the same thing as one of our upper limbs. This is, actually, the exact definition paleontologist and comparative anatomist Richard Owen gave for "homologous" in 1843—for the first time connecting the term to those structural relationships that were already discussed in the works of Belon and Geoffroy Saint-Hilaire. As previously mentioned, for Owen *homologue* is *"the same organ in different animals under every variety of form and function."* This definition should clearly be placed in a pre-evolutionary cultural context. The comparison between one animal and the other, or the organ of one and that of the other is developed by Owen within a framework devoid of historical dimensions.

It is not difficult, however, to reinterpret this notion of homology by adding to it precisely this sense of living beings' becoming in the course of evolutionary events. Homology thus becomes the correspondence that exists between the structures of two different organisms to the extent that they can be traced back to the same structure of a common ancestor: more precisely to the structure of their *most recent* common ancestor. The forelimb of a frog or the upper limb of a human being, for example, are homologues of one another since they derive from the upper limb of one of the first terrestrial vertebrates, starting from which the evolutionary lines that led to amphibians and modern mammals respectively began to diverge.

This historical reinterpretation of the notion of homology has, in the course of the last fifty years, been further elaborated within so-called cladistic systematics. This is a methodological approach that has profoundly renewed and transformed the manner in which biological systematists operate by employing a rigorous comparative analysis of characteristics that should allow for an ever more accurate and complete reconstruction of the kinship relations between different species.

But the fate of the concept of homology has not been exhausted by its reinterpretation in a historical vein or by its application in the reconstruction of the phylogenetic tree of living organisms. Many researchers, among those interested in developmental biology especially, have shown an increasing dissatisfaction with an interpretation of homology in terms of derivation from *common ancestors*, and have instead begun searching for *common mechanisms* that are shared by the developmental processes by means of which homologous structures come to be realized. Given what we have previously stated about the relations between genes, development, and organic forms, it is significant that this effort to base the analysis of homology on developmental mechanisms rather than on history has quickly led to an awareness that the simple tracing of two homologous structures to the common possession of this or that gene would not have been an adequate explanation for the correspondence between the two structures.

One Gene, Several Effects

The principal reason not to accept such an identification between shared genetic expression and the homology of organs developed by two different animals lies in a phenomenon known as pleiotropy. Pleiotropy asserts that many characteristics are influenced, within the same organism, by the expression of one and the same gene. Naturally this concept is endowed with precise meaning only if we are able to accurately define what a gene and a characteristic are. And in fact, both terms are problematic, and even though biology continues to make extraordinary progress in these fields, the ambiguity or inadequacy of these and other concepts persists. We discussed this issue briefly in the preceding chapter.

Here I will for once adopt a pragmatic solution and ask the reader to think of the gene, without too many caveats, as a segment of DNA that contains information for the synthesis of a protein, and to think of a characteristic as any morphological aspect that an observer can "pick out" from an accurate description of an animal (or a plant). This said, we are ready to face the countless examples of pleiotropy that developmental genetics is continuously uncovering.

Generally speaking the proteins codified by single genes perform specific functions for a cell's vital processes, but these functions are often shared by cellular types that differ greatly from one another.

Consequently the same gene, by means of the protein it codifies, is often involved in the construction or functioning of very different organs. This is somewhat analogous to what occurs in our residences. The electrician's work, for instance, is indispensable for bringing electrical current not only to lightbulbs, but also to the washing machine, the iron and the microwave, and to putting us in touch with the outer world via telephone or computer. And this is only one of the network of relations that exists between specialized technicians and the structural components of our house. The bathtub and maybe even the sink need the work of a bricklayer, and perhaps even of a tile-layer. Neither one nor the other, however, will be concerned only with the tub and the sink, but they will apply their work to many other structures that only they are competent to handle. Each of these technicians, therefore, leaves a trace that we might define as "pleiotropic" on many "characteristics" of our house, in the same fashion in which the expression of a gene has consequences for a multiplicity of structural and functional aspects in a living organism.

The issue becomes more complex, naturally, if we also take into account the fact that genes are not those unitary, monolithic structures that perhaps we would prefer to deal with. Ultimately a gene is not only made up of that strand of DNA, which, when translated into amino acids according to the rules of the genetic code, precisely corresponds to a protein. From a functional point of view at least, several other strands of DNA also belong to the gene, and the circumstances, the places, the timing, and the manner in which the gene itself will be expressed depend on them.

Therefore, when we compare the structures of two different animals and the genes that seem to be involved in their production, can we state with certainty that they are homologous since they are controlled by genes that contain the information for basically identical proteins in the two animals? In other words, can we simply ignore the differences that may exist between the two species in the non-codifying portion of the gene, in the part that is not reflected in the structure of the protein, but has a role in deciding when and where the gene will be expressed?

Ars Combinatoria

In my opinion these sorts of questions lead to one necessary conclusion: the abandonment of the traditional notion of homology as a relationship that either does or does not exist between two structures. Instead we should, for each comparison, highlight the particular aspect according to which we want to establish the presence of a relation of homology between the two species. That is, the time has come to make room for a factorial, or combinatorial, concept of homology. This in turn will mean that the two structures being compared may be homologous according to certain criteria but not others. More specifically two organs may be homologous because of position, but not because of their specific qualities, or vice versa.

I previously mentioned the *Drosophila* with a little eye on one leg obtained experimentally by Walter Gehring and his collaborators in the late 1990s. The microscopic structure of this eye is precisely that of the normal composite eye of a *Drosophila* (and extremely complicated, it is), with a great variety of well-differentiated cellular types arranged according to a precise geometry. Certainly this is an eye that is incapable of functioning, since it is deprived of those necessary and sophisticated connections that a sensory organ must have with the brain; but this lack does not deprive it of its specific characteristics as an eye, which, on a structural level, is absolutely comparable with a normal eye. There are therefore many reasons for stating that this "accessory eye" is homologous to a normal eye, but only from the point of view of its internal structure and its cellular types; and this is of course already an important insight. It remains, however, an *ectopic* eye, in other words an eye that is in the "wrong" place, not equivalent to that in which a *Drosophila*'s (or any other insect's) eyes are normally formed. Therefore: yes to the special homology (relating to the organ's structure), no to the homology of position. This distinction, which once would have been considered a vacuous or arbitrary verbal exercise, in fact possesses absolute legitimacy, since the relative *position* of the different structures within the animal are rigorously controlled by a series of genes that are expressed in a precocious phase of development. The zootype's genes, which we have previously mentioned, are their best-known component.

The connection between a specific position and a specific structure that undergoes differentiation there tends to be preserved in the course of evolutionary history, but it is not a necessary and indissoluble connection. A good example of this inertia, and, simultaneously, of the possibility of sometimes breaking an ancient connection between position and structure, is represented by the genital orifices of arthropods—apertures that most probably were originally two (one on each side), but that in many cases (for instance, in most insects) are reduced to only one, on the ventral aspect of the animal.

In this case, however, we are not interested in number, but in position. The location of these orifices is about halfway down the body in many groups of arthropods, for example in scorpions and crabs, but it has moved elsewhere in at least three groups: millipedes, centipedes, and insects. In the first group it has moved forward, to the level of the second pair of legs; in the others, it has moved backward: in centipedes to a literally terminal position, and in insects to the level of the third to last segment. Notwithstanding these variations, the genital aperture obviously maintains its nature, its appearance.

What has occurred in the course of the evolution of these structures can be figuratively described as follows: originally the genital aperture was fixated to a "positional marker" close to the body's midpoint, but in some groups of arthropods it became detached from this marker, was repositioned, and was attached to markers in various positions, closer to the front end (millipedes) or the rear end (centipedes and insects).

This reconstruction leads to some interesting questions, and one of them back to our initial query about the relationship between homology of position and special homology. We can in fact ask ourselves if, in those arthropods whose genital aperture is no longer at the body's midpoint, the corresponding "positional marker" has been moved or abolished. It could, instead, have remained in place, but now serve as a marker for structures other than the genital aperture.

Rigorous research on the issue, at the molecular level, has yet to be undertaken. There are however good morphological clues that lead one to think that the "marker" has survived in its original position. New and different structures, present in each of the three groups in which the genital aperture moved from its original position, seem in fact to be connected to this position.

In the males of many millipedes, one pair of legs, at the level of what would have been the location of the original genital aperture, has been transformed into gonopods, appendages that help the animal hold the female during insemination. At about this same level, the regularity with which the many segments of the centipede's body appear to follow one another, seems, in many species, to be interrupted: it can, for instance, affect the otherwise undisturbed sequence of segments that alternatively possess respiratory apertures or not. Finally, in the males of dragonflies, on the ventral aspect of the second abdominal segment, short but complex appendages are present, which, while not directly connected to the genital aperture, in any case function as copulatory structures, thanks to which the transfer of spermatozoa to the female occurs. All these structures would therefore be in a relationship of positional homology with the genital aperture of crabs and scorpions, even though they have nothing in common with them from the point of view of special homology. The position of each of these structures relative to the main body axis of the respective animal is probably the same in all these cases, not only (or not so much) in metric terms, in other words in terms of the relative distance from the body's two extremities, but also, in all probability, because specified by the expression of the same combination of "position marker" genes in the respective embryos.

The Limits of Hierarchies

Homology can therefore be insidious terrain, as one often enters the traditional blind alley of the all-or-nothing approach, according to which the homological relation between the two structures compared either does or does not exist. This is the result of a shortsighted perspective that is incapable of discussing the criteria according to which we select the specific characteristics compared in the phenotype. These are characteristics that may correspond to easily recognizable anatomical-functional units, such as the hand or the heart, but don't have an equivalent in the uniqueness of specific developmental processes. Referring to the genotype instead of the phenotype has not proven to be a very useful strategy either, because this path easily leads us to the quicksands of pleiotropy. It is therefore necessary to find the necessary analytical criteria by means of which the complexity of the phenotypic characteristics being compared are reduced to a set of more easily comparable aspects.

But we should not delude ourselves that the problem can be solved by following the usual analytical procedures via hierarchization. Though the human being's upper limb may seem too complex, it is not sufficient to consider the arm, forearm, and hand separately. And if the hand is too complex, it is not sufficient to separate the palm from the fingers and, continuing in this manner, analyze the fingers' phalanges separately. In fact, each of these parts makes us confront a structural complexity that in its turn leads to a set of more or less distinct morphogenetic phenomena. These phenomena, moreover, are widely shared by all these subunits and give rise to the bones, nerves, blood vessels, and so forth that we find in each of these subunits. Therefore if we limit ourselves to this hierarchical analysis, we would be in the situation, to continue the earlier analogy, of calling the electrician separately for each room, each bulb, each meter of electrical wire that had to be laid at our house. And the same would go for the painter, the plumber, and the tile-layer. We need all of them on the ground floor and on the higher floors and on each floor, in each room.

Certainly from many points of view the rooms in our house are interesting subunits, but sometimes it is more interesting to examine the set of all windows, or all radiators or all electrical outlets. Or that of the tables or the chairs in the house. Or that of the possible bed spaces we can rapidly set up should relatives suddenly come to visit whom we don't feel like sending to a hotel.

All these "things" (windows, radiators, outlets; but also tables and chairs; and even potential things, such as the bed spaces we just mentioned) are part of our house or its décor just as much as our rooms are. But the different partitions intersect one another in many ways, and they do not just follow a simple scheme of hierarchical inclusion. This scheme might, perhaps, be valid for windows, each of which is cut in one wall of a room and never two; but the same cannot be said for a chair, which we can move at our whim from one room to the next. It would also not apply to electrical wire that, in addition to being laid from one end of the house to the next, can constantly or temporarily provide current to many different appliances.

As in the case of our house, and perhaps even more so in the case of the analysis of a living organism, we must be aware that this operation can be performed in many different ways. Some of these are certainly more obvious, such as the distinction between trunk and appendages, or the division of a finger into phalanges, but these are not necessarily the most interesting paths. In each case the criteria for the analysis of an organic form, and, consequently, the aspects according to which we are about to perform a comparison between two or more forms, are neither given a priori nor are they exhausted by the schemata suggested by anatomical textbooks. This awareness therefore obliges us to abandon the facile guidance provided by our old manuals, and it opens up the endless horizons of a world of forms where homology cannot but be, as we have said, factorial and combinatorial.

Organs, Only When All Is Done

Such a widening of horizons may instill a sense of uncertainty in researchers. If they have to abandon the old patterns, from where shall they start again? The problem certainly does not consist of finding some "pieces" that can be the legitimate object of comparative exercises. On the contrary, because the field is now open to a wide spectrum of legitimate criteria for comparison, we can pick and choose. And it is here that the difficulties arise. Since it will never be possible to undertake an exhaustive analysis of highly structured systems such as living organisms, what should we focus our attention on, in the hope of gathering a booty of interesting results? I don't think this question has any optimal answer. But I do think we can put the experiences that have taken place during these years in the area of evolutionary developmental biology to good use and take two insights with us.

The first concerns the identification of parts we would think of as operative units in our comparisons. This, basically, is the question I briefly examined in the previous chapter. The reader will certainly remember the increasingly serious uncertainties that surround the traditional concepts of segment and tissue. We should underscore, however, that these uncertainties do not affect all the possible applications of a concept, but demonstrate the risk one runs when a concept originally used within a specific context is then employed in other contexts.

Thus, it is best to be prepared for surprises, even big ones. Let's think for instance about our concepts of hand, foot, liver, heart, stomach, and brain. Each of these terms corresponds to a part of our body with a very specific structure that, in the economy of the organism to which it belongs, exercises a set of functions that is equally well defined. It is therefore reasonable to think that each of these organs, following the Darwinian paradigm of evolutionary biology, must have reached its current structural and functional characteristics under the pressure of a selection that affected countless generations endowed with forms that were more or less similar to those these same organs have today. And it is just as reasonable to think that each of these will be realized, in the course of individual development, through a series of specific stages, under the control of a long series of genes that progressively enter the game as ontogeny proceeds. But this is precisely where a trap exists. If in anatomical terms (and also, to a certain extent, in physiological terms), it seems fairly easy to "pick out" a liver, a heart, or a brain as distinct parts of an animal, are we sure that the same is true from the point of view of development? In other words, are there distinct processes, basically autonomous in relation to the rest of development, that one could designate hepatogenesis, cerebrogenesis, and cardiogenesis? In my opinion, these questions must receive a negative answer, even if the set of developmental processes that end with the realization of the liver, the brain, or the heart are in many respects different from the set of processes that lead to the genesis of other parts of the body.

Each part of the body is characterized by its own special combination of expressed genes and cellular events—most of which are shared (but with different types of association) by many other parts of the same organism—rather than by the exclusive expression of a certain number of genes or the appearance of morphogenetic processes that are not shared by other developing organs. To speak more accurately, a list of the truly distinct morphogenetic events includes phenomena such as cellular migration, or the organization of a group of cells to form a sphere or a tube by virtue of the properties of mutual cohesion and adhesion to other cellular types. Organogenesis, however does not belong here since it represents a convenient category in which we place some of those showy temporal segments of an intertwined set of elementary morphogenetic phenomena operating in the same body district that end with the appearance of something that will have the "dignity" of an organ, finally defined in an anatomical and functional sense. From the point of view of development, perhaps, organogenesis has never possessed such an autonomy and individuality.

Structures, Processes, and Developmental Stages

The second insight that comparative biology can glean from evolutionary developmental biology is, in a certain sense, an extension of the preceding comment, except that it can be made by those who are still convinced that it is worthwhile to talk about specific types of organogenesis (hepatogenesis, cerebrogenesis, etc.). This insight, in fact,

opens up a whole new category of comparisons: not between anatomical structures this time, but between processes—even metabolic
ones, naturally—above all between those with which we are most
closely concerned, that is to say developmental processes.

From this point of view, there is not much difference between
asking if the segments of arthropods are homologous to those of
annelids and if there is a homology between the segmentation processes of the two groups. In passing from one formulation to the
other, however, the issues we actually focus our attention on do
change. In the first case we will continue to be interested in structural aspects and, above all, in the relationships of reciprocal position between anatomical parts, while in the second case, these aspects will recede toward a distant horizon, and the timing and
manner in which this or that gene expresses itself, or the manner in
which certain cell populations dialogue (or do not) with one another, will come to the fore instead.

Paying closer attention to these dynamic aspects can finally lead us
to further types of comparison, in this case no longer between single
anatomical structures, or between elementary processes, but between
the temporal intervals in which these processes are realized—in
other words, a comparison between stages of development, for instance, between one larva and another. If the temporal or causal
bonds between the different developmental processes that become
intertwined in the course of an animal's life can be cut and, perhaps,
tied up again in a different manner, it is possible that this exercise in
comparisons between stages will be as problematic as a comparison
between structures. And then this might require the adoption of a
factorial, combinatorial criterion. It is possible for example, that
stage A of animal X may correspond, in certain respects to stage A,
and in other respects to stage B of another animal Y. Let us try for
instance to compare the olm, or *Proteus*, the blind salamander of the
caves of the Dinaric karst, with one of our common newts. In the
phase in which it reproduces, the olm is comparable to an adult newt
to the extent that it is sexually mature; but the olm still breathes by
means of gills, like the newt's larva, and not with lungs, like the adult
newt.

This possibility of associating, dissociating, and reassociating different aspects in the development of an organism can be of great importance from an evolutionary perspective. And it is precisely in these aspects of *heterochrony* that one can find an important key to the understanding of the origin of evolutionary innovations. We shall discuss these issues again in the last chapter.

Chapter 9

The Body's Syntax

From Tip to Toe

I mentioned the body's syntax in chapter 4, but only cursorily. In an evo-devo context, however, this is not a minor issue, and I shall therefore look at it more closely here. We can start by revisiting what are usually called the body's axes.

Let us take a grasshopper, for example. There should be no problems in identifying its longitudinal axis (the body's main axis), along which the head, the thorax, and the abdomen succeed one another. The longitudinal axis of an earthworm is equally obvious, even if in this animal there is no clearly distinct head, and the entire body appears to be composed of a long series of similar segments.

From a certain point of view, things are even simpler in an earthworm than in a grasshopper. In the earthworm, in fact, both the mouth and the anus are located at the extremities of what we identify as the body's main axis, while their arrangement is less linear in the grasshopper whose mouth is directed downward. Shall we therefore say that the longitudinal axis of this insect is not rectilinear, but has a 90° bend at the front, or shall we acknowledge that this axis, which passes by the anus at the rear, ends with the forehead instead of the mouth at the front (fig. 10)? A choice between these alternatives is ultimately a simple question of conventions. Underlying this issue, however, there is a fairly interesting biological problem. To construct an almost cylindrical animal with mouth and anus at the two extremities seems to be simpler than constructing an elongated animal with the mouth directed downward. It is not only a problem of

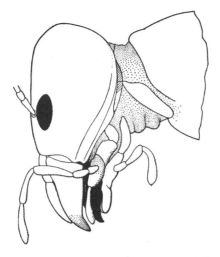

Figure 10. Insect head with the mouth oriented toward the substrate.

adaptation (it is certainly convenient for the grasshopper to have a mouth with an opening directly above the leaves it can eat), but also a problem of morphogenesis. Here too, then, we have a situation that simultaneously involves evolutionary biology and developmental biology.

Naturally the issue does not only concern the grasshopper and other insects whose mouth is also directed downward. Other and perhaps greater complications surround the body's main axis. From the earthworm, for instance, we can pass to a close relative, the leech. In this case the complication affects the rear part of the body and is related to the development of a sucker. Truth be told leeches actually have two suckers, but the one in the front surrounds the mouth, which basically maintains a terminal position as in the earthworm. At the animal's other end, however, things are different. If the rear sucker were to be found at the very end, and the anus opened up within it, the leech would often be faced with a dilemma: defecate, and thus risk loosing contact with the substrate, or remain attached, thus postponing defecation to an uncertain and indeterminate later moment. In actual fact the sucker represents the specialization of seven posterior segments and it is therefore really terminal, but the anus migrated forward and opens dorsally, immediately before the sucker itself. This ruse solves one of the leech's existential problems, but it creates a problem for the zoologist who wants to describe the main body axis of this annelid in an adequate fashion. The situation is similar to the one we encountered in the case of the grasshopper, except that the deviation from the rectilinear condition is, this time, close to the rear end, and is directed toward the back rather than the belly. But this is just the beginning. A considerably more difficult (i.e., more interesting) case is that of the sipunculus.

The Dual Animal

Since we are discussing an animal that is not very well known, I would advise the reader to become familiar with its appearance before continuing (fig. 11). It is a marine animal, not especially rare on sandy beaches, that can occasionally be found beached, especially after a storm at the end of winter. The sipunculus is an animal with a tapered, almost cylindrical body; therefore recognizing its body's principal axis should not constitute a problem. And in fact its external appearance does not present any problems. But there is a complication. The sipunculus's digestive tract is not straight like that of an earthworm. Starting with the mouth, easily recognizable at one of the tapered body's two extremities, it continues until it almost reaches the body's opposite extremity, but here it folds in on itself, and even twists around the preceding straight tract, moving forward once again until it opens, with the anus, at a point not far from the mouth. Does it therefore make sense to define the body's longitudinal axis solely (or particularly) on the basis of the animal's external appearance?

Actually one could propose a different solution. Instead of using one descriptive grid referring to the animal's exterior appearance, it might make sense to distinguish the two components, which, as we have seen, seem to be somewhat independent of each other. For simplicity's sake, we could talk of "exterior animal" and "interior animal." If we want to use more technical terms, we can translate these terms into "somatic animal" and "visceral animal," respectively, but the real-

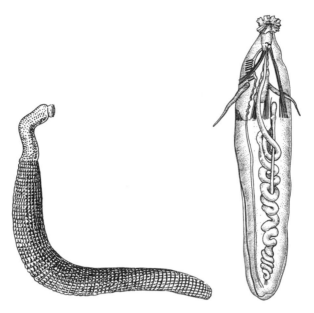

Figure 11. Sipunculus.

ity remains the same. The exterior animal is that "sack" principally comprised of the epidermis and the underlying musculature, which contains the interior organs and hides their curious arrangement, at least for an observer not interested in pursuing dissection. Its longitudinal axis is the one we intuitively identified in our worm at first glance. The interior animal, however consists of its inner organs, most especially the digestive tract. Its longitudinal axis is the axis of the intestine, from mouth to anus. That it is not straight is related to the fact that this visceral animal is contained in a shorter somatic animal.

When the two "animals" are the same length, as in the earthworm, there are no problems. Or, better, the situation is such that we do not realize that the somatic and visceral animals are independent of each other, not only from the point of view of the functions they perform, but also from the point of view of the developmental processes of which they are a product. Generally we begin to become aware of the issue only when the animal does not exhibit an anteroposterior axis, as in the case of numerous sedentary animals such as sea squirts, although it is sufficient to think of the manner in which our intestine is lodged in our belly.

The absence of an alignment between somatic and visceral animal, therefore, instead of representing an obstacle for a standardized description that refers to the usual body axes, opens new and interesting opportunities for research. In the course of development, how is the regular alignment of the different parts of the somatic animal realized, from one extremity to the other? How is the regular alignment of the different parts of the visceral animal, from the mouth to the anus, realized? And, finally, what reciprocal influences between the two sets become manifest during the different phases of their differentiation? These are problems that developmental biology has only marginally begun to confront. And it is clear that they could not emerge until developmental biology became aware of the importance of the comparative method that evolutionary biology had adopted long ago.

The Sea Urchin

Since we are discussing body axes, an even more serious problem is represented by sea urchins. In order to adequately understand the nature of the question, we summarize here the principal phases in the development of these animals. A regular embryo is formed from the egg, which initially exhibits radial symmetry and in its turn is soon transformed into a larva (the pluteus) that exhibits bilateral symmetry. Up to this point nothing strange. But inside the pluteus, there is the outline of a future adult that soon assumes a regular five-ray radial symmetry. For many sea urchins, this is the final stage. There is, however, a group of species known by the (somewhat exaggerated) name of irregular sea urchins. The extent of their irregularity consists in their not sharing the radial symmetry of their kin: during metamorphosis, they resume bilateral symmetry. The mouth still opens onto the lower side, as in their "regular" counterparts, but the anal opening is moved backward compared to its usual position, which corresponds to the center of the upper side.

Let us leave the irregulars aside, and ask what the main body axis of a sea urchin with radial symmetry might be. What trace of the longitudinal axis of the larva is left? It is probable that the entire architecture of the adult's body has been realized *ex novo*, without utilizing larval geometry as a point of reference. If this is the case, then maybe we should ask if regular sea urchins really have only a single bodily axis—because maybe they have five, all equivalent to one another. An accurate study of their metamorphosis and of the different calcareous plates that enclose the sea urchin's body suggest that, in actuality, one plane of symmetry is slightly different from the other four, which intersect it along the axis that goes from the center of the mouth to the center of the anal opening; but as to what conclusions to draw from these comparisons, echinoderm experts are still debating. While we await a better understanding of this issue, let us shift our attention to an animal whose appearance is very different: the tapeworm.

The Tapeworm

If we got stuck in our attempts to interpret a globular form with nearly perfect radial symmetry like the sea urchin, we might think that the tapelike form of a tapeworm shouldn't excessively trouble us: but reality is not on our side. This time the problem is not identifying the animal's main body axis, which we can take for granted, but rather finding out which is the front and which the rear end. Two unfavorable circumstances make this determination difficult.

First, the tapeworm does not move. If it did, we would tend to consider the extremity that advances first as the front end, which, precisely because it is involved in exploring the surrounding space, would probably be equipped with sensory organs connected to a nerve center (a brain, let's say), perhaps located nearby.

Second, the tapeworm does not have a mouth; to be more precise, it has no trace of a digestive tract. If there were a mouth at either extremity, we would almost certainly say that this is the front end. We should note, however, that in planarians (nonparasitic animals common in rivers that belong to the zoological grouping of the flatworms, just like tapeworms) the mouth opens onto the ventral side, in a fairly backward position, closer to the rear end than to the front. Planarians, however, are mobile animals and it is because of their movement and the concentration of sensory and nervous cells at one of the body's two extremities that we can determine its polarity in a (perhaps) non-ambiguous manner. We should also specify that the anus is always absent in flatworms. But let us return to the tapeworm.

In the absence of the two most obvious clues for determining its polarity, one could ask if the overall shape of the animal is of any help. And in fact the two extremities are very different from each other. At one of the ends is the scolex, a structure equipped with suckers or hooks (or both suckers and hooks) that allows the tapeworm to attach itself to the wall of the digestive tract of its host. The rest of the body is made up of a succession of units, the proglottids (whose shape and number vary a lot from one species of tapeworm to the next), which are continuously generated, in the larger species at least, in a region that is very close to the scolex. The oldest proglottid, therefore, if it has not already detached itself, is the one furthest from the scolex.

All zoology books describe the scolex as the tapeworm's front end. Opposing interpretations of the polarity of this parasitic worm's body were actually suggested by a German researcher and an American zoologist at the beginning of the twentieth century, but they were immediately discarded. And yet there are at least two good reasons to suspect that the scolex is the tapeworm's rear end, one of which was postulated in two "heretical" articles by Ludwig Cohn and Emma Watson.

If we interpret the tapeworm's structure according to the current paradigm, we in fact encounter an unusual, not to say paradoxical situation. Generally, in the course of development, animals elongate starting from a posterior, terminal, or subterminal proliferative region. During the embryonic development of a vertebrate the anterior vertebrae are the first to be formed, followed, in that order, by those progressively closer to the coccyx. And the same goes for the segments of a leech. It remains true that some invertebrates are also capable, when necessary, of regenerating the anterior part of the body, even if less easily than the posterior, but this fact does not constitute a sufficiently strong argument regarding the difficulty that has been raised in the case of tapeworms. Why should they elongate starting from a region that is very close to the body's front end, something that other animals don't do? The second objection is more subtle and requires a brief but important aside, and a small (but in my opinion convincing) exercise in comparison.

This aside concerns an aspect of the body's syntax that has been given little attention to date. It is a question of the position, relative to the animal's anteroposterior axis, of the animal's male reproductive organs as compared to its female ones, and their respective openings. An alignment of these organs such as to allow an evaluation of the positions they occupy is not always possible, and, in any case, is often a delicate operation in animals with distinct sexes. This problem, however, does not exist in the case of hermaphroditic animals. In this latter case, in fact, male and female organs are present in the same individual, even if they do not always mature contemporaneously. In those cases in which an alignment is possible, the female organs mostly occupy a position that is closer to the animal's front end. In planaria, those nonparasitic flatworms mentioned previously, the two ovaries are found very far forward, almost at the level of the eyes, while the testicles have a much more backward position.

In the case of tapeworms, we must consider a single proglottid (fig. 12). In each of these units, one can find a complete genital apparatus, both male and female. And the relative disposition of testicles and ovaries in the proglottid is the opposite of what one would expect, if one follows the principle that the scolex represents the front end of these worms' bodies. Naturally things fall into place if we adopt the opposite interpretation, with the scolex at the rear end.

Doubt as to the legitimacy and the extension of the argument may seem valid because the comparison was made between tapeworms and animals like planarians, which are flatworms that differ greatly from them. The scenario would be even worse should we involve animals from different and remote zoological phyla. But if we search inside the grouping of flatworms, we can find a more suitable

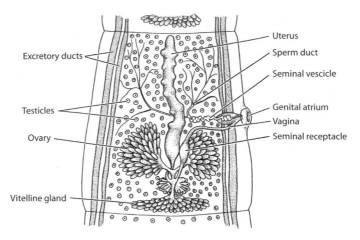

Excretory ducts

Testicles

Ovary

Vitelline gland

Uterus

Sperm duct

Seminal vescicle

Genital atrium

Vagina

Seminal receptacle

Figure 12. Tapeworm proglottid.

comparison. These are the monogeneans (fig. 13), a group of parasitic flatworms that have two great merits in regard to the comparison we are about to undertake. First, in terms of kinship these are the animals closest to tapeworms (using a technical expression, we could say that they are tapeworms' *sister group*), and like them they are hermaphroditic. Second, they are equipped with a mouth situated at one of the body's two extremities, and this easily allows us to determine the polarity of monogeneans, notwithstanding the fact that these animals, like tapeworms, are also completely sedentary and cannot therefore give us any clue as to the orientation of their main axis relative to the direction of their movements.

In monogeneans the ovaries are anterior relative to the testicles, thus conforming to the empirical generalization we formulated above. At this point I think there is nothing left for us to do but to . . . upend all the tapeworm illustrations in our books and await a final verdict that will perhaps be forthcoming once the problem has been adequately examined with the tools of developmental molecular genetics.

As we already know, numerous genes are expressed along the anteroposterior axis of animals' bodies following a sequence that is widely shared. Among these are the *Hox* genes. But *Hox* genes are

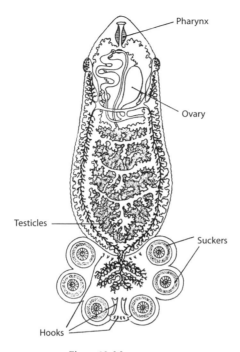

Figure 13. Monogenean.

probably not the best suited to resolve the problem of the tapeworms' polarity, because, generally, their expression does not extend to the entire body axis, as the most anterior section is usually excluded. Here, however, other genes are expressed, such as *Omx* and *Emx*, which have mostly been studied in vertebrates, but are much more widely distributed. I hope that some laboratory will soon confront the issue and that we will be able to find two or three genes whose spatial sequencing will provide a perhaps definitive answer to the question of the tapeworms' polarity.

PART THREE

Origins

Chapter 10

Competition or Cooperation?

Apologues and Metaphors

In an ancient Rome where social groups had a hard time finding stable and robust forms of social integration, Menenius Agrippa addressed a famous apologue to the secessionist plebs in which he compared the different components of the republican state to the different parts of the human body. The functional unity of the biological individual was used as an unchallengeable example of an integrated system in which the parts cooperate for the benefit of the whole.

For others, such as Thomas Hobbes in *Leviathan*, life is a *bellum omnium contra omnes*, and it is not difficult to see the same principle at work in Charles Darwin's *Origin*, with all the consequences that had for an acceptance of evolutionary theory by free-marketeers, anarchists, free-thinkers, and those tied to religious beliefs. There are those who interpret human history as an inextinguishable succession of conflicts between individuals, peoples, ideas, and institutions and those, like Arnold Toynbee, who imagine that our species' path leads toward one great civilization, finally stabilized in time, that will make use of the best that past civilizations had to offer.

Competition or cooperation? Undoubtedly the events of history, natural and human, have many examples to offer in support of each view. It is therefore not necessary to look for them outside of biology. Let us examine some that have a direct bearing for the themes we are discussing.

The Species: From Inside and from Outside

At the risk of moving, even if briefly, to the extreme periphery of the topics we are dealing with in this book, we should examine two opposing perspectives regarding one of the most important (and problematic) concepts in evolutionary biology, the biological concept of species—one focusing on cooperation, the other on competition.

By way of brief introduction, the biological concept of species is based on the reproductive compatibility between individuals and therefore on their capability to give birth, through breeding, to offspring who in their turn will be able to perpetuate the characteristics of the species.

There are, however, as stated above, two principal versions of this concept: one follows the logic of competition, the other the logic of cooperation. The first version underscores the existence of a reproductive barrier between species. The male of a mouse and the female of a rat, or a hamster, will never produce offspring, as mouse, rat and hamster are three different species. A male donkey and a mare, however, can give birth to a mule, and there are both male and female mules, but if they mate, no offspring is born. The reproductive barrier that exists between the donkey and the horse is temporarily overcome, if human beings break down the natural condition of isolation, but the attempt is exhausted after the first generation. Therefore horse and donkey are, and remain, distinct species.

For individuals belonging to species A all the individuals belonging to all the other species are therefore, from the point of view of reproduction, quite indifferent, while those that belong to their own species are a resource (if of the opposite sex) or competitors (if of their own sex). From this point of view it is possible to define the species as the most inclusive set of individuals who are able to compete with one another for access to the same set of reproductive resources. This definition may appear curt, since it is expressed in the realistic language of economics. It has, however, met with a very favorable response in biology (in zoology more than botany), even if its practical application is not always easy and sometimes actually impossible. But this issue would take us too far from the subject of this book. Let us instead see if the situation presented by two close species and their reciprocal relationships can be described in a language that differs from that of competition for access to reproductive resources.

An alternative description, in terms of cooperation (or sharing) rather than competition (or exclusion) has been proposed by Hugh Paterson. This entomologist of South African origin has underscored an already well-known fact, but one that had not been previously utilized: the "signature" of each species. In other words, the fact that members of each species share a recognition system that allows them to correctly identify members of their own species (especially those of the opposite sex), thus ensuring the establishment of "legitimate" couples and preventing a significant portion of the reproductive effort being wasted in fertilization attempts between gametes that are not, or are barely, mutually compatible. These recognition systems include the songs of birds, of frogs, of crickets; the flashes of light emitted by fireflies; the nuptial parades of the peacock; the chemical signals released in the water by the eggs of many marine animals and capable of steering the movement of spermatozoa to their source; and the pheromones by means of which female moths attract male moths, even from great distances.

Paterson has suggested grouping all these types of signals exchanged between individuals of the same species under the category *specific mate recognition systems* (with its acronym, SMRS). Plants may also have their SMRS, it is just that the signals they emit are not received directly by members of the same species, but are gathered by animals (above all bees, butterflies, or other insects) that visit their flowers. The specificity of the flowers' shapes, colors, and perfumes guarantees that the insect will identify them with precision. This helps explain the faithfulness with which many insects visit flowers of the same species, at least for a certain period of time. The result, naturally, is the transfer of pollen from one flower of a species to another—that is to say, ultimately, the successful sexual reproduction of the plant.

One can therefore see that the biological concept of species, the notion that it consists of a potential or actual reproductive community, can be interpreted in two different ways: by *internalizing,* which highlights cooperation and the sharing of reproductive resources, or by *externalizing,* which highlights competition for those same resources.

The species is not the only structural level for which evolutionary biology allows this dual interpretation. We could say the same things, for instance, about ants belonging to the same nest. It is true that each individual ant has its needs in terms of food, space, and so forth, and is therefore in competition with the other members of the community to which it belongs. But this community is in its turn in competition with the communities of the other nests, and at this level of analysis the ants from each nest represent an integrated unit, cooperating with one another in a fashion that is advantageous for the whole community. In the following paragraphs we shall look at some examples of competition and cooperation that concern developmental biology more than evolutionary biology. At the end of the chapter we shall attempt to draw some overall conclusions from the examples given.

Butterflies and Sea Urchins

In the course of their brief existence, many insects undergo profound metamorphoses. Let us take the butterfly, for instance. To tell its story in a few lines I will begin with the egg, following a well-established tradition.

Out of the egg, then, comes a little animal with an elongated, almost cylindrical body equipped with numerous pairs of appendages. This is the caterpillar, and during the course of several weeks, it will generally undergo four or five larval stages separated by molts, and then, maybe after having enclosed itself in a papery cocoon, it will be transformed into a chrysalis. This is an apparently quiescent stage, whose immobility actually hides a profound internal transformation, both structural and functional. Many of its internal structures are in fact demolished as the adult insect takes shape: a butterfly with a slightly shabby appearance, but with its four wings, its legs (always six, like those already present on the caterpillar's thorax, but much longer and more clearly articulated than these), and the characteristic parts of its mouth (which are no longer those of the caterpillar, made for chewing leaves, but rather the typical retractable straw, thanks to which butterflies drink their liquid food). The butterfly is additionally equipped with a fully developed reproductive apparatus, both in its internal structures (ovaries or testicles) and in its external ones, which it will be able to utilize for copulation and ovipositing. These structures were not present in the caterpillar and are built during metamorphosis.

The transformations that many marine invertebrates undergo are no less "catastrophic." As I wrote in a previous chapter, the embryonic development of a sponge, a sea urchin, a starfish, an oyster, and many other denizens of salty waters does not end with the formation of a miniature copy of the parent, but with the formation of a larva, and it is almost impossible to guess the identity of the future adult from the larva's appearance alone.

Given the radically different organizational plans that separate these larvae from their respective adult phases, we should not be too surprised if they were initially described as separate genera and given names (trocophora, mitraria, pelagosphaera, etc.) by which they are still known today. Names that today are written in lower case, to underscore the fact that they are not distinct genera, but instead simple stages of development on a par with those that have been named, for instance, Higgins' larva or Müller's larva, in honor of those who discovered them.

The metamorphoses that these larvae undergo are, in some cases, even more catastrophic than those (already impressive) that affected the butterfly's caterpillar or *Drosophila's* larva. In the case of the sea urchin, the larva (known as a pluteus) is a small gelatinous and transparent thing exhibiting bilateral symmetry (and not radial symmetry, as will be the case, more or less exactly, with the adult). The larvae contain a little group of cells that for a while seems to remain at rest, excluded from the vital functions (locomotion, nutrition, interactive life) of the larvae, but that at a certain point begins to grow, soon revealing the characteristic five-ray symmetry characteristic of the sea urchin. This predecessor of the future adult grows rapidly, to the detriment of the larval structures, which are soon reduced to a residue that is destined to disappear completely. One could almost say that the adult has used the larva like a parasite (or, better, a parasitoid) uses its victim.

Parasitoids

A parasite (whether a tick that attaches itself to the host to suck its blood, or a tapeworm that lodges in the digestive tract to utilize some predigested food) sooner or later harms its victim, but does not kill it. This is no indication on its part of either sadism or temperance; it is instead a sign of the importance that prolonging the life of the host has for the parasite itself. There are cases, however, in which the victim is quite literally consumed within the period of time it takes for all the nutritional requirements of the aggressor to be met. At this point, it is too late for the victim to hope to recover, and the enemy no longer has any interest in keeping it alive. In a certain sense, here, we are halfway between predation, which implies the killing of the victim, followed by the predator's meal, and true parasitism, in which, as we have said, the life of the victim is prolonged notwithstanding the damages the parasitic attack inflicts.

This intermediate strategy is precisely the one undertaken by parasitoids, of which there are thousands of examples in the world of insects, especially among hymenopterans. The females of many species of this order, in fact, lay their eggs next to the eggs or larvae of another insect or, more often, inside. Once hatched, the larva of the hymenopteron will grow, nourishing itself exclusively with the victim its mother assigned to it. Once the meal is concluded, nothing will be left of the victim but an empty skin, while the larva of the hymenopteron will be ready to change into a pupa (the quiescent stage that in butterflies is given the name chrysalis) and then into an adult.

Throw-Away Larvae

Let us return to the sea urchin. From a certain point of view the fate of its larva, progressively resorbed by the rudiment of the future adult, very much resembles the fate of the victims of parasitoid hymenopterans, but there is a profound difference between the two situations. In the case of the sea urchin (but this is certainly not the case in the other story), victim and exploiter have the same genetic information. It is as if one part of our body were to grow by utilizing another part, which slowly atrophied until it disappeared. Let us imagine a brain, for instance, that developed by absorbing matter from a leg that thus became progressively shorter, or from a foot that thus became a simple rudiment or even completely disappeared. Please note that we are not talking about an evolutionary tendency that over the course of numerous generations sees one part of the body developing ever more, while other parts follow an opposite tendency. This is the case, for instance, with the progressive growth of the cranium, and the brain it contains, in the evolutionary line that led to what we currently know as human beings. This tendency is accompanied, in the same evolutionary line, by the disappearance of the tail, whose skeletal trace is reduced to a minuscule remnant, the coccyx. In the case of the sea urchin, the opposition of larval organs and adult organs is a true competition, an internal struggle between cell groups that derive from a single fertilized egg, and, as we have said, therefore share the same genetic information.

Similar developmental histories, which contemplate the elimination of an extensive portion of the body of the larva, are repeated in other zoological groups. Let us examine two more examples, the nemertines and the digeneans.

Nemertines are wormlike animals, almost all marine, with elongated bodies. Most species start their life among the plankton, in the guise of a diminutive gelatinous larva with the shape of a top, which is then transformed into a cylindrical worm that slowly moves along the ocean floor. The larvae of nemertines are not all the same, but for our purposes, it is sufficient to follow the transformations of the most widespread model, the pilidium. One can identify six points on the surface of this larva starting from which the epidermis is introflected, forming six pockets. A seventh pocket is formed, however, through the extroflection of the intestinal area. These seven pockets gradually expand until they join one another and the epidermis closes again, forming an external film, beneath which there is a space filled only with liquid that contains the internal cell mass. It is precisely from these pockets that the adult develops, while the external pellicule, a larval residue, is soon completely eliminated.

We now move on to digeneans, or flukes (fig. 14). In this case we are talking about parasitic flatworms, with a complicated biological cycle. Normally their existence alternates between two very different hosts, a vertebrate and a mollusk. The vertebrate can be either aquatic or terrestrial, while the mollusk almost always belongs to an aquatic species, even when the vertebrate host is terrestrial. It is easy to imagine that the transfer from one host to the other is relatively easy when the mollusk and the vertebrate are both aquatic, while the passage of the same parasite from a terrestrial host to an aquatic one or vice versa would require special devices.

In this instance we are not interested in delving into the topic of the evolution of parasitism, fascinating as it may be. We are instead talking about larvae, and it will be sufficient for us to focus on a part

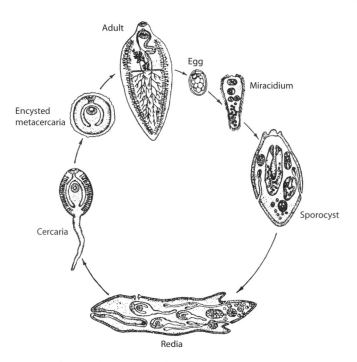

Figure 14. The life cycle of a digenean.

of the biological cycle of one of those digeneans that never need to leave the water. The worm that is hosted by an aquatic vertebrate produces eggs, and from each of these a larva emerges that is called a miracidium. This larva swims until it finds a mollusk, and once inside, it is transformed into a worm with a shape very different from its initial one. This worm then reproduces in its turn, generating larvae of a still different type, which are called cercariae. These move in the water, and, having sooner or later reached an aquatic vertebrate, they develop into an adult similar to the one from which the story started.

Let us now place the miracidium under a microscope. It is covered by a series of large flat cells, equipped with cilia, which allow the little larva to move in the water. At the front end of the miracidium is a structure (terebratorium) that will allow it to penetrate inside the mollusk. Underneath the terebratorium is a small brain to which sensory cells are connected that help the larva search for the host. Finally there are some excretory cells. Apparently there is nothing else. More specifically there is no mouth, no digestive tract, and no reproductive organs, either. At bottom the miracidium is only a transitory stage, an exploratory phase. Only after penetrating the new host will there be an opportunity to obtain food, to grow and reproduce. But the worm that installs itself in the mollusk no longer has the shape or the structural organization of the miracidium. The large ciliated cells that once enveloped the larva are literally abandoned as soon as it is transformed into a sporocyst, whose body develops from a large cell mass that was located in the miracidium's rear end and is covered by a new epithelium that had already been forming in the larva underneath the large ciliated cells.

The complex metamorphoses that these larvae undergo raise many questions, but I will only raise two here. These are, however, formidable issues and cover the whole spectrum of problems this book attempts to deal with. The first issue is of an evolutionary nature. The problem is in the possible historical reconstruction of events that have led to the present biological cycles of sea urchins, nemertines, and digeneans. Must we in fact imagine that their larvae (at least those that can move freely like the sea urchin's pluteus or the nemertines' pilidium) resemble the adults of very remote ancestors of these modern forms, while the current adult forms represent innovations that progressively appeared in the course of evolution? If this is not the case, should we instead believe that these larvae represent innovations, the result of a detour from the path that originally led directly from the egg to an adult that was basically similar to a sea urchin or a nemertine? Until now zoologists were almost unanimous in believing that these larvae were copies of the remote ancestors of the present forms whose adults would instead represent innovations that progressively evolved over the course of centuries. But I believe that the problem must be reconsidered with a mind devoid of prejudices, like the one that leads to the all too facile equation simple equals primitive, and from which comes the conviction that the larvae, generally less complex than the respective adults, are a heritage of the extremely remote past. We should rather expect natural selection to be quite conservative as far as the adult's structure is concerned. It is in fact normally the adult that is engaged in reproduction, while natural selection could be less rigid as far as other stages of development are concerned, so long as these innovations do not entail negative consequences for the organization and the reproductive capacity of the adult. Ultimately in the case of insects with a complete metamorphic cycle like butterflies and beetles, it is reasonable to state that the larvae are more "innovative," in evolutionary terms, than the adults.

Competition between Equals, or Not Quite

The other issue concerns developmental biology above all. The problem, in fact, consists of understanding in what manner the two groups of cells that originated from the same fertilized egg could initiate a competition so strenuous that it leads not only to the resorption of one group by the other, but actually the physical removal of a group of cells, whose possibility of survival is abruptly cut short.

Let us therefore address the problem of competition between groups of cells that share the same inherited genetic information.

This last clause—the sharing of the same genetic information—must be taken literally for two reasons.

The first has to do with issues that occurred a long time ago in the history of the Earth. In fact it concerns the appearance of the first multicellular organisms. There is no doubt that this event was one of the most critical transitions in the entire evolution of living forms. In fact, if it is true that the transition to a multicellular status was only possible for some groups of unicellular organisms, it is also true that it offered extraordinary possibilities for further evolution. These possibilities are right before our eyes and, ultimately, the main subject of this book. As far as the conditions that allowed for the realization of nonoccasional forms of multicellular organization, we will focus here on a single problem: the kinship relation between the single cells that constitute a unit made of many cells.

To better familiarize ourselves with the nature of the issue, we shall do well to distance ourselves for a moment from the manner in which things take place in almost all multicellular organisms. In order to construct one such organism, we can in fact imagine a method that is really different from the progressive series of cellular divisions that start from an egg, a seed, or a spore. This method consists of the aggregation of preexisting cells that led an independent life before aggregating. Even if this method of constructing a multicellular organism differs from the one adopted by animals, plants, and mushrooms, one cannot state that it has never been tried by living organisms. On the contrary we have an example that is readily available, not in a fossil state, but in a small group of colonial amoebas that live among dead leaves. Precisely because of the peculiar manner in which it passes from the unicellular to the multicellular condition, one of these amoebas (*Dictyostelium discoideum*) has become an important model organism used in many laboratories.

So long as environmental conditions are good, *Dictyostelium* leads its normal unicellular existence feeding on bacteria and, occasionally, reproducing by means of simple cellular division. But things change once conditions become more difficult—specifically, when food starts to be scarce. Attracted to one another by a chemical signal that they produce, amoebas join to form an ever growing heap. At a certain point, once a critical mass has been reached, they all begin to move together, almost as if they are a diminutive snail. Their voyage does not last long and ends once some of the cells that formed the little migrating flock stop, while others clamber on top of them until they form a sort of stem or antenna. Still more cells move to the apex of this structure and ultimately form a spherical mass. These last cells then form a protective coating and transform into spores. Even a very slight mechanical stimulus will be sufficient to dislodge them, dispersing them for some distance around—perhaps only a few millimeters, but this will be sufficient to carry them to new "pastures" rich in bacteria, where they will be able to shed their protective coating and once again become active amoebas (fig. 15).

Once we abandon these spores to their fate, we can focus for a moment on the numerous cells that will not share this destiny, because they are instead designed to perish after having served as a launching pad for the others. These are the cells that formed the

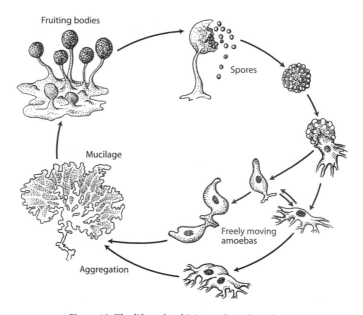

Figure 15. The life cycle of *Dictyostelium discoideum*.

diminutive antenna, and on whose extremity the mass of spores was formed. It is only the latter spores that have hope for a future, which is instead definitively denied to the other cells.

Where is the problem, one will ask. Even in the case of an animal, future prospects differ from one cell to the next. In most cases one can draw a clear demarcation between those cells that will be able to contribute to the formation of a new generation and those that are instead definitely excluded from this opportunity. The first group are the cells that belong to the germ line, the ones that will give rise to the gametes (eggs and spermatozoa); the others are the somatic cells, those that form the skin, the brain, the liver, the heart, and the other organs, and that will disappear without a trace once the individual dies. One can indeed say that the cells of the somatic line are destined to die, while those of the germ line have some hope of immortality by way of ensuing generations. This entire account, naturally, contains various exaggerations.

The most obvious is the one regarding immortality (all things considered, rather improbable, even for the best-endowed gametes). The second exaggeration, and one almost as obvious, consists in the fact that a very large number of gametes (above all, among male specimens, but usually also among female ones) will in any case have no future, since the available space on our planet is finite (and already abundantly occupied). The third and least obvious one is that in some cases not all cells of the somatic line are completely denied the opportunity of a future beyond the end of the mortal life of the individual of which they are a part. This occurs very often in plants, less often in animals. This last phenomenon is described in general biology treatises as agamic or vegetative reproduction. Plants produced from a cutting and polyps from a parent's bud are examples of this form of reproduction. And let us observe that in plants the decision about which cells will be able to become ovules (and then seeds), or give life to pollen granules, instead of forming leaves, branches, or petals is taken very late. In the young plant, while it is still growing, the options are still open; in fact, the game has not yet even begun.

Germ and Soma

But let us return to animals where the distinction between somatic and germ line is taken early on: very often at the onset of development. In some cases these opposite destinies are actually dictated by the mother. This occurs in those cases in which, during the maturing process inside the ovary, a specific structure is formed in the cytoplasm of the egg, the so-called germinal determinant. Later on, once the egg has been fertilized and begun its divisions, the matter that makes up the germinal determinant will be divided between a small number of cells, the ancestors of future gametes, while others, which will not share even a trace of the determinant, will be left to follow the fate of somatic cells.

On the whole, however, this distinction between somatic and germ line seems fairly legitimate and sensible. *Non omnis moriar*, something of me will remain, in future generations. Having said this, how can we not accept that everything else should share the common fate of mortals? Let us observe, moreover, that the germ line itself would not have a future were it not aided, day after day, by the cells of the somatic line. Even the sack that contains the future gametes (in other words, the gonad, testicle, or ovary) is formed by cells from the somatic line, and the same goes for the canals that allow the gametes to emerge into the open, not to mention the copulatory organs, or the ovipositing organs, which are those structures by means of which the females of many animals (insects, for example) place their eggs in appropriate locations, often inside some fairly firm substrate. And it is certainly not the gametes that form the rooster's comb or the gaudy livery of the male birds of paradise. Just as it is not thanks to the cells of the germ line that a little bird will engage in its trills, or that a stag will engage a rival in an exchange of head butts. In a certain sense, therefore, gametes use gonads, muscles, the brain, feathers, or antlers in the same fashion as the spores of *Dictyostelium* use the cells of the stem that help take them higher. Is there no problem, then? No, competition between cells is a serious issue, and we had therefore better be clear about who belongs to which line: who belongs to the germinal line and who belongs to the somatic line, who to the little globule of spores and who to the stem that takes the spores higher and then dies. And this is precisely where the difference between animals and *Dictyostelium* lies.

In the case of animals, human beings included, all the cells, whether they belong to the germ or the somatic line, derive from a shared ancestor (the fertilized egg), and they therefore share inherited genetic information. If there is competition between them, it is a competition between equals. Naturally we are not talking about ethical issues here. What matters is that, whichever the cell is that will have a future in the next generation—among the many that have descended from the same fertilized egg—it will take with it a copy of the same genetic information that is contained in any other cell belonging to the same individual. In the case of *Dictyostelium*, however, the many amoebas that gather to form a little flock that will later be transformed into a spore mass (stem plus spores) don't all descend from a common recent ancestor. To be more precise, such a degree of kinship is possible, but is not guaranteed. Certainly, given the modest speed with which these amoebas move one could expect that among those spatially closest to any given amoeba that will eventually join the growing multicellular mass, there will mostly be cells with the same genetic information.

Dictyostelium amoebas reproduce by means of simple and repeated cellular divisions, the same process that in animals leads to the formation of a growing number of cells starting from a fertilized egg. But the cells that derive from a fertilized egg remain together, while the two amoebas that form from the division of a preexisting amoeba separate and start their journey through the wide world, each minding its own business. They can then, sooner or later, gather with amoebas they are not closely related to. Perhaps, at this point, one of those famines that induce them to aggregate with their neighbors begins, a situation in which they cannot find out if any of these other amoebas are related to them or not.

And this is precisely where the danger lies. In a flock formed by amoebas with differing genetic information, there can be a type of cell that for some reason is more often capable than other amoebas of reaching the top of the stem, so as then to give rise to spores. Just a tiny difference in the ability to adhere to the other cells, or in the relative speed of movement within the fluid mass that the little flock represents, could be sufficient. If this occurs, there will necessarily be exploiters and the exploited. This distinction is legitimate since those cells that are transformed into spores and those that instead constitute the stem are not identical copies of one another, but differ precisely because of a hereditary characteristic that ensures some of them will achieve a success that is denied the others.

It is therefore not difficult to understand why there are very few living organisms in which a multicellular state is reached by way of aggregation, as in *Dictyostelium discoideum*, while in all other cases a different, genealogical, path is followed, in which all the cells of the unit derive from the same progenitor cell. In the first case in fact there is no guarantee that these cells, now gathered together, are genetically identical. Therefore a slight opening remains for the possibility of competition between one strain that is capable of exploiting another strain, and a second strain that risks being exploited. This instability does not exist when all the cells have the same genetic information.

We have therefore verified the first of the two reasons I mentioned earlier that invite us to take literally the clause regarding the identical genetic makeup of cells belonging to a multicellular unit. It is only in this case that it seems possible to avoid a competition that either in the shorter or the longer period would guarantee an advantage to only one of the parties involved, given, naturally, that it manages to perpetuate itself without the aid of the other, a problem that does not seem to exist among the known strains of *Dictyostelium*.

The other reason lies in the fact that a cell line in which a mutation develops within the multicellular organism of which it is a part can sooner or later put the survival of the individual at risk. It could, for example, simply be a question of a mutation that stimulates cellular proliferation, removing a cell (and its descendents, whose offspring are obviously destined to grow rapidly) from those forms of control that would otherwise keep it quiescent. As a result of this mutation, a genetic difference, minuscule but with significant consequences, would have been introduced between the cells of this tumoral line and all the other cells of the organism. Since their advantage consists precisely in their reproductive capacity, the tumoral cells end up being the winners in the competition with the rest of the organism. In an evolutionary perspective, however, this victory appears, dramatically, as a Pyrrhic one. The death of the individual in which the rebel line has appeared inevitably also affects this line's destiny. In any case this second argument also confirms the profound difference that exists between a competition among genetically identical cells and a competition among cells that differ from one another even (or above all?) only in one gene.

Authorized Competition among Equals

Once we have clarified the conditions that make a competition between somatic line and germ line acceptable (in other words sustainable, from an evolutionary perspective), it is perhaps less difficult to conceive of the nature (if not the origin) of the competition that undoubtedly matches the transitory structures of the larva against those, victorious in the long term, of the adult, during the development of a sea urchin or a nemertine. We should rather ask ourselves if these facts represent anything other than a particularly visible expression of a more widespread situation, where the competition, during the course of its development, between cells, tissues, or organs belonging to one and the same individual, would represent the norm: inevitable perhaps, and perhaps by now "assimilated" by natural selection.

In fact, the competition between different parts of the same developing individual is a universal phenomenon. But it is difficult to notice as long as one remains tied to two traditional ideas—on the one hand, the conviction that what matters especially in biology (and should therefore be the principal object of our studies) is the individual as a whole, and therefore its constitutive parts are seen in light of their contribution to its development and well being; on the other hand the conviction that development exists only with the purpose of building the adult individual. Apoptosis, in other words "programmed" cellular death, is therefore welcome if it is necessary for the individual's well-being. Without the death of some rows of cells in the early rudiment of our limbs, we would find a membrane between our fingers similar to that found in waterfowl. There is no reason to lament the death of those cells, or to exalt the success of the neighboring cells that are saved from apoptosis. What matters is the construction of a limb suited to our purposes. Selection will reward it, thus also rewarding the apoptotic mechanism that made its construction possible.

But this interpretation, as satisfying as it may appear from the point of view of adaptation, in other words of evolutionary biology, is not as satisfying from the point of view of the origin and evolution of developmental mechanisms. To understand these, an interpretation in which the individual is the protagonist, rather than the single cell, cannot take us far. A finalistic interpretation of development has even led in recent years to the paradoxical statement that apoptosis is the most likely fate of any animal cell, a fate from which it can escape only if surrounding cells send it the appropriate signals to save it from an otherwise unavoidable end.

I am convinced that such a view represents an inverted view of reality. By analogy, we should then in fact state that the fate of every lemming is to jump into a fjord and drown (naturally before it has reproduced!), unless sudden drowsiness allows it to rest in its den on the day on which almost all its fellow lemmings leave on their desperate journey. Apart from the fact (and I beg forgiveness from ethologists for having introduced this example) that the lemmings' situation is less tragic than was once depicted, it is impossible to deny that a life based on suicide does not contemplate great future prospects. It therefore seems improbable that the developmental mechanisms of the entire animal kingdom would be based on it. It instead seems reasonable to interpret the phenomena tied to apoptosis (which are controlled by precise molecular mechanisms, partly shared by animals that belong to distant zoological lineages) as one of the many forms that competition between cells endowed with an identical genetic information can assume within that multicellular system that we designate as an animal in the course of development.

Moreover this competition is not necessarily so brutal as to lead to the physical elimination of one of the contenders. It is, in fact, a competition for resources (food above all) and, therefore, the possibility of maintaining a sustained metabolism, of growing and reproducing.

A Virtual Mouth

To realize just how universal such a competition is, it is sufficient to consider the earliest stages of development, starting from the fertilized egg. These first stages consist of the cleavage, that is to say the subdivision of the initial egg mass (or one of its parts in cases in which the egg is large and full of a yolk that does not participate in these divisions) into an increasingly larger number of cells. In some cases (in nematodes, for instance), already at the stage in which there are two cells, one notices differences between them. In other cases, one has to wait for more divisions to occur, but it is unlikely that the process will continue with perfectly symmetrical and synchronized divisions when the egg has already subdivided into several dozen parts. Sooner or later all embryos exhibit differences between one cell and the other, as regards both their size and the time it takes for them to start the next division. Only in its earliest stages, then, is the embryo constituted by a number of cells equivalent to an integer power of two. As the number of cells that form the embryo grows, it in any case becomes impossible to ensure that they all have an identical environment. It is in fact easy to imagine how eight identical cells can be arranged to form a sort of cube with two cells per side, but as the number of cells grows further, in order to remain identical to one another, they would all have to assume the shape of a pyramid with its vertex at the center of a sphere, or, in order to preserve a less improbable shape, they would have to dispose themselves so as to form a thin layer (only one cell thick) around a cavity, either empty or filled with liquid. This shape is actually an approximation of the embryonic blastula stage that can be identified in the early stages of development of many animals. But a group of cells joined together by a sufficiently strong adhesive force will in any case tend to form a full mass, whatever their number. Sooner or later, then, a contest between the cells that remain on the surface, in contact with the external world, and those that are instead covered by the surface cells, emerges.

This is a contest that was "accepted" and stabilized in remote periods of evolutionary history, giving rise to two mutually differentiated cell populations capable of coexisting in a state of reasonable equilibrium. In the simplest cases this state of affairs is realized by the internal cells disposing themselves so as to form a second layer that adheres internally to the first. However we must add an important clause: the internal cells also gain access to the external world, thus creating an opening that will serve as a mouth or, perhaps, as an anus. But this only becomes possible by "negotiating" relationships between the inner and the outer leaflets. Sooner or later the border between the two must be fixed. How this took place, in the earliest animals equipped with two leaflets, we do not know. Perhaps, however, we will soon be able to conceive of this process by following a precious clue that came from research conducted on the hydra about ten years ago.

The hydra—we mentioned it earlier—is a small freshwater polyp whose body, practically speaking, is simply an elongated little sack formed by two layers of cells precisely juxtaposed. The bottom of this sack is attached to the substrate. At the opposite end there is the mouth, surrounded by a crown of long, thin tentacles. To tell the truth, the mouth exists and it doesn't. It is there, and it opens when it is time to ingest a morsel—a diminutive crustacean or other small prey—but soon after, it seems to literally disappear. It we take a hydra that has just recently ingested its victim, treat its tissues with a fixative, cut it up into little slices, and observe it under the microscope, we won't find its mouth. More than sealed: vanished. And yet, a while later, the mouth can reappear at exactly the point at which it had opened previously. Some trace must therefore have remained in the meantime. A molecular trace, probably, that corresponds to the thin ring where the opening (real or virtual) of the exterior leaflet of the little sack is attached to the opening (real or virtual) of the interior leaflet. And, in fact, this trace not only exists, but seems to be represented by more than one molecule, the most important, however, probably being the protein codified by the *Brachyury* gene, which has its equivalent in basically all animals, including mammals. The border between the two leaflets is no longer in contention.

Embryonic Leaflets

Two leaflets, one of which (the ectoderm) forms the body's exterior wall, while the other (the endoderm) forms the wall of the digestive tract, are more or less clearly identifiable in the embryos of all animals, with the exception of sponges (fig. 16).

In a creature as simple as a hydra, this duality of leaflets and the structures formed by them represents almost the entire anatomy of the little polyp. In almost all other animals, however, the situation is more complicated. According to the classic descriptions in embryology textbooks, the mesoderm, the third leaflet, intervenes (literally) as well, and its derivatives don't face the external world, but are enclosed between the two original leaflets. Our muscles and skeleton originate from the mesoderm, for instance.

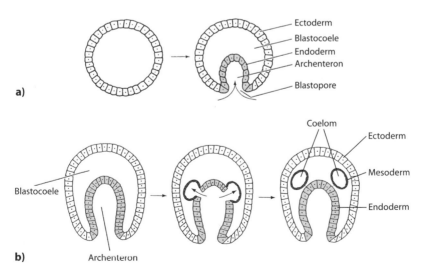

Figure 16. Embryonic leaflets: (a) diblastic condition, (b) triblastic condition.

With the appearance of the mesoderm we are no longer confronted with a two-player game in the embryo but—at the very least—with a three-player game. In actuality, the number of distinct cell populations that will share the resources—large or small, but in any case limited—the embryo has at its disposal is much higher, even though tradition has wanted to especially underscore this distinction between two or three groups of cells that have the characteristic of developing early and of being easily recognizable even in distant zoological lineages. It should not surprise us therefore if some researchers have also elevated other groups of cells to the "dignity" of being distinct embryonic leaflets.

At the end of the last century, Brian Hall proposed attributing the title of "fourth embryonic leaflet" to a group of cells that is characteristic of vertebrates and is endowed with a distinct identity, an unusual behavior and a precise role in the realization of important anatomical parts. This cell group is the neural crest, a set of cells that is located in a dorsal position relative to the neural tube (the rudiment of the main axis of the nervous system: brain and spinal cord), but is destined to migrate far away, toward well-defined locations where it will give rise to varied series of tissues and other different structures—among these, elements of the sensory and the autonomous nervous systems, the medullar portion of the suprarenal glands, various pigmented cells, and even a part of the head's skeleton.

Why not accept this proposal by Hall? Even if its population acquires its individuality relatively late compared to the traditional three embryonic leaflets, there is no reason to not grant the cells of the neural crest equal "dignity." The problem lies elsewhere. If we want to be fair we should also grant many other groups of cells in different animals the same "dignity." This is the case, for example, of the cells that form the imaginal disks of insects with a complete metamorphic cycle like *Drosophila*, bees, and butterflies: cells whose population is sorted out even before the beginning of larval life and do not participate in the construction of larval organs. For a long time these cells limit themselves to increasing in number, up to the moment in which, during the pupal stage, they are the main actors in some of the most outstanding events in the metamorphosis of the insect: the construction of the antennae, the eyes, the wings, the legs, and the external genital structures of the adult insect.

One might object that the imaginal disks never form a unitary leaflet or mass. From their first appearance, they develop as a set of distinct units located at different points in the larva, which match the places where the corresponding organs in the adult will develop. But this is not a serious objection. Do the cells of the neural crest form a compact population? On the contrary, they are divided into a series of small groups, each of which migrates along a different path and results in the differentiation of distinct cellular types that become part of different structures in the adult individual. And, to move to a different example, is the United States not a unitary nation simply because it includes Alaska and Hawaii?

Having said this, why deny the title of distinct leaflet to that set of cells which forms pockets in the larvae of nemertines from which the adult worm will develop?

The list is, naturally, longer, but unnecessary in our discussion of the problem we began with, the competition between distinct groups of cells within the same animal. In both nemertines and insects with a complete metamorphic cycle, the adult is formed at the expense of the larva, while salvaging some parts (the nervous system above all else) and redeploying everything else according to a different structural plan.

This occurs, to repeat, within the confines of a system that is constituted of cells that all have the same genetic information, and we should therefore expect the same situation in the offspring, the same competition between individual cells or leaflets or, in any case, between parts or stages—whatever name we want to use—of the same animal.

Who Is Winning?

Naturally, these forms of competition are particularly obvious when the animal is in a stage in which it doesn't eat but lives, for a while, exclusively at the expense of those resources it already disposes of. Some insects, for example, don't partake of any nourishment during the entire course of their adult existence. Everything they eat derives from the nourishment they accumulated during their juvenile existence. Among these are mayflies, insects with a delicate and fragile body that pass most of their life in the water, eating algae, detritus, and small prey, depending on the species in question, only to then be transformed into adults with an atrophic oral system and therefore incapable of eating.

In these conditions a dramatic competition arises for the limited resources available. The insect could prolong its active life by many days by using these resources mostly for the relatively modest needs of its basal metabolism and for those, much more costly, of the contraction of its muscles in flight. In this fashion however, it would leave little or nothing for the gametes it might produce. The eggs, above all, are very costly because of their quantity and the energetic value of the yolk that is accumulated in them. But if all the resources they accumulated during their juvenile existence were spent on gametes, a noneating adult would not even have enough breath to move one millimeter from the position in which it finally opened its wings. Germ or soma, therefore: which will prevail?

It is clear that the only acceptable solution in evolutionary terms (i.e., the only solution that will not lead to the immediate extinction of the species) is one that contemplates some form of compromise between the two contenders. In other words the insect should be able to mature its gametes, but must also be able to survive and busy itself at least enough to ensure that its eggs, or its spermatozoa, will be able to meet their complementary gametes. In the case of ephemerids (mayflies), the balance of the competition between germ and soma is heavily on the side of germ, and so the life of the adult insect is extremely short: often a single day, or even less, as their name indicates.

The eggs and the embryos that derive from them are generally closed systems (with some exceptions that we won't deal with here) in that they don't take in nourishment from external sources. This is also true of the pupae of insects with a complete metamorphic cycle.

I referred to the events that occur inside an embryo while discussing embryonic leaflets, but it will be worth our while to also mention some peculiar (and outstanding) situations that generally extend beyond embryonic development into more advanced stages of the animal's development.

One Egg, Several Embryos

We are talking about polyembryony—literally, the formation of multiple embryos. One fundamental caveat should be assumed: these many embryos start from one fertilized egg. It is not a widespread situation in nature, but it does concern our species. Polyembryony is the condition that gives rise to two or more identical individuals. Under what heading shall we classify it? Is it a special reproductive modality or a peculiarity of developmental processes?

As often occurs in the natural sciences, when exploring polyembryony, we face a phenomenon that exposes the insufficiency, or the excessive rigidity of our classifications. If reproduction is any phenomenon that results in an increase in the number of individuals in a population (a closed population, obviously, so as to be able to discount the phenomenon of immigration), polyembryony certainly deserves to be included in this group. But it is also true that, at an early stage, starting from the moment the fertilized egg begins its divisions, polyembryony involves an important detour from the ordinary sequence of stages an embryo passes through.

Setting aside issues of semantics or classification, polyembryony is an interesting phenomenon in at least two respects. First, one must believe that the subdivision of the matter that initially constituted the egg, or the little mass of cells that derived from it after several divisions into several embryos, will destroy the geometry the embryo exhibited up to that point. If the embryo already had some body axes (a dorsal and a ventral side, a front and a rear end), it is probable that not much would remain. Second, polyembryony offers us an ideal situation in which to study competitive phenomena. And this too deserves our attention.

The two or more embryos that derive from the single fertilized egg must share the available resources. This situation is not serious in the case of a mammal such as a human being in which the egg contributes only in a negligible manner to the energy demands of the embryo because the transfer of nutrients by the mother soon comes into play. The competition that develops between two monovular twins therefore tends to resemble what can exist between two twins who derive from two distinct fertilized egg cells.

In the two cases the limits imposed by the available resources are equivalent. One difference, naturally, is the existence of genetic identity between the two embryos (or fetuses), or the lack thereof. When they are not identical, one can expect that one will be more efficient than the other in taking control of the available resources. There should therefore be no adaptive reasons against the appearance of polyembryony. Moreover, polyembryony is the rule with the nine-banded armadillo, which always gives birth to four identical individuals. If polyembryony is not more frequent than it actually is, the problem perhaps lies elsewhere: in the probability that it will evolve, rather than in its adaptive value.

Two Twins, or Rather One

It is a fact, however, that in the few known cases outside of mammals, polyembryony is associated with a series of other rather unusual phenomena. Let us look closely at two of these.

The first concerns a genus of little South American freshwater fish (*Cynolebias*) that belongs to the group of so-called annual fish, who acquired this name because of the brevity of their biological cycle. The first stages of their embryonic development seem to foretell a case of polyembryony. And in actual fact the two cells that are produced by the first division of the fertilized egg separate and give rise to two independent small cell masses that proceed with their divisions and begin differentiation, first of all giving life to an exterior and interior leaflet. Before the completion of this important step, however, the two embryos fuse back together and soon no recognizable trace of their temporary condition of duality remains. A competition between not very ferocious rivals, or ones that are too similar to one another (they still descend from the same fertilized egg) to continue warring? Unfortunately at the moment it is not easy to substitute these metaphors suited to the human condition with reasonable causal explanations. Let us, however, note this phenomenon and examine a different situation wherein the initial polyembryony is no longer in doubt—as in the fish of the genus *Cynolebias*—but is instead translated into a dramatic difference in development and, as a final result, into two subgroups of identical embryos. Yes, subgroups, because in this case the identical individuals that derive from the same egg are not two, or four, but several hundred. Up to two thousand embryos.

Identical Twins, or Maybe Not

In this second case, we move from vertebrates to insects, to a polyembryony that involves some species of diminutive parasitoid wasps. Here it will suffice to examine only one, a tiny insect whose females deposit their eggs in some of the eggs laid by moths.

The discovery of the polyembryony of this insect is due to an Italian researcher, Filippo Silvestri, who, at the beginning of the twentieth century, was one of the pioneers of biological pest control, the technique used to contain insects that are noxious to cultivated plants by fielding their natural enemies and, more specifically, the parasitoids that attack their eggs or larvae. Silvestri called the little wasp in which he discovered polyembryony *Litomastix truncatellus*, whereas today specialists think it is more appropriate to call it *Copidosoma truncatellum*. (I mention this fundamentally irrelevant nomenclature issue because readers might encounter both names and think that different animals are being referred to).

In the last several years, studies on *Copidosoma* have become fashionable again, and they are being undertaken with the most sophisticated techniques of developmental molecular genetics; however, these labs are no longer working with Silvestri's old species, but with an American species that belongs to the same genus, *Copidosoma floridanum.* At present, no differences are discernible between the two, at least as far as polyembryony is concerned. Contrary to expectations, in recent years both Silvestri's observations and his interpretations, which had been previously questioned, have been proven correct. Here, briefly, are the facts.

A single egg of *Copidosoma* is deposited, as I said, inside a moth egg, and a caterpillar develops from the moth egg, notwithstanding the presence of a mortal enemy within it. A caterpillar that has been attacked by *Copidosoma* in such a fashion can reach even larger dimensions than a healthy caterpillar of the same species. Compared to the little enemy wasp, it is therefore a huge victim. In fact, one single caterpillar can fulfill all the nutritional needs of several hundred *Copidosoma* larvae! No harm would be done to the caterpillar if the wasp deposited only one egg in its victim: the wasp's solution, naturally, lies in polyembryony—one egg, but several embryos, that is, more larvae and, in due course, more adults.

The problem is that the same moth egg could be (or could have been) also stung by another parasitoid in addition to our *Copidosoma*. And if this were to happen, the competition for food would no longer be restricted to the more or less numerous sister larvae that are derived by polyembryony from the single *Copidosoma* egg; and in this case, among its adversaries, there would be the larvae of a different species of wasp. From the point of view of a larva of *Copidosoma*, which must share its food with both its identical siblings and extraneous larvae, what it must avoid at all costs is succumbing (or the risk of succumbing) to this extraneous presence. Having to move aside to allow an identical twin to survive would, all things considered, be much less serious.

What occurs inside a moth egg in which a female of *Copidosoma* has deposited its egg, or in the caterpillar eventually deriving from it, pretty much fits our expectations. Initially the wasp's egg gives rise to a single compact mass of cells, approximately equal in size and corresponding to an early embryonic stage that is traditionally designated morula because it looks like a little fruit from a mulberry tree (whose Latin name is *Morus*). We are still at the beginning, before the cells that constitute the embryo differentiate into distinct germ layers, and everything still looks as if a single larva will develop from the fertilized egg. But this is not the case, and suddenly the morula fragments into a large number of diminutive units, each formed by a few cells. This is polyembryony. However, unlike what occurs in other cases, as in human monovular twins or in the little armadillo quadruplets, the larvae that develop from these diminutive partial morulae are not all identical to one another, and are instead clearly divided into two classes. Some have a normal appearance, comparable to those of the larvae of other wasps. Barring unforeseen accidents, each of these will complete its development, ultimately giving rise to an adult. The others, though, have a thin wormlike body, and their internal structure is simplified. These larvae will never complete their metamorphoses. If we could describe their fate in a language applicable to human beings, we would say that their lives will not be spent in vain. They are, in fact, soldier larvae, ready to attack and kill extraneous larvae of other species should they be found inside the same moth larva. In this manner they clear the field for their normal sisters, facilitating their survival and, thus, increasing their likelihood of completing development. An unusual case of altruism, one might say, that in some ways resembles the behavior of soldiers in societies of bees, wasps, and ants. These soldiers are also (save rare exceptions) precluded from the hope of reproduction, but they undertake actions that are indispensable to the reproductive success of other individuals (a queen sister in this case) to whom they are closely related.

In the case of *Copidosoma,* the degree of kinship between the sterile soldier larvae and their fertile sisters is the highest possible since they belong to an identical group of monovular individuals. All the cells of all the larvae that descend from the same egg naturally have the same genetic information. This is a relationship similar to the one we encountered between the cells of that single embryo that in most animals derives from a fertilized egg.

From a strictly genetic point of view (obviously not from the morphological or functional ones), a set of monovular twins is like a large embryo divided into two autonomous districts. And it is precisely because of this autonomy that we would expect less competition between these monovular individuals than the one that can exist between two cells from each of them, or from any other embryo. Except naturally for the fact that the monovular twins will in any case compete for access to the resources available in their environment. If the environment is small and closed, like what is inside the corion (i.e., the shell) of a moth egg, inevitably this competition will be pronounced. And not only for food, but also—if at all possible—as a defense against extraneous presences.

But with the separation of the *Copidosoma* larvae into two distinct castes, the sterile and the fertile, we are not only confronted with a case that resembles that of the social insects, but in some sense are facing the primordial antagonism between somatic and germ line. This comparison, as we shall see immediately, can be taken almost literally.

In insects, and therefore also in *Copidosoma*, the location of a corpuscle (the oosome), formed inside the egg before it is fertilized, will decide which cells become part of the germ line. In a certain sense, then, this is an issue that is "decided" by the mother. When, after fertilization, the egg begins to divide, the matter that the oosome is made of is included in a small number of cells, which as a result become the ancestors of the germ line. In *Copidosoma* the oosome soon fragments into numerous particles, each of which is sufficient to confer the value of ancestor of a future germ line to a cell, but in the fragmentation of the initial morula (in other words when polyembryony becomes manifest), not all the offspring morulae receive an oosome fragment. Those larvae without a germ line will derive from those morulae that remain without any oosome. In this insect, therefore, polyembryony is associated with an unusual division between germ and somatic lines. Many larvae will partake of both, but some will only be part of the somatic line.

In a finalistic view of developmental processes, we could comment on these events as follows: since they are deprived of a germ line, it would be a waste to allow these larvae to complete their development. I think it is more valid and appropriate, however, to underscore the elegance with which, in this competition between cells, the germ line of these insects manages to "manipulate" the behavior of the somatic line in both its expressions: the soma of the normal larvae, where it lives together with the germ, and also the soma of sterile larvae. But we remain within a set of genetically identical cells, none of which, as a matter of course, should overpower the others.

The Pupa's Balance Sheet

Certainly, it is easy to be generous when one is wealthy. When resources are limited, however, competition comes to the fore. From this point of view, cells find themselves in a special condition during the pupal stage.

We have mentioned that the profound structural transformations that occur within the pupa are often so radical that they completely deprive the animal of any possibility of active movement, at least for a while. Some pupae, for example those of many moths, conserve a certain flexibility in their abdominal segments; those of the ant lion are actually able to move their mandibles, but in any case the pupa does not ingest food.

Even not taking the basic cost of its metabolism into consideration (though there are no expenditures due to muscular activity for locomotion, it still has to continue to breathe), the pupa must reckon with a balance sheet that does not foresee any income beyond its initial capital. Metabolic expenses, the unavoidable rejects, and the part that cannot be mobilized, which corresponds to those structures that pass virtually unscathed through the process of metamorphosis, are subtracted from this capital. The rest will be divided between the new structures: the wings, the antennae, and so forth, that are characteristic of the adult.

Many of these structures, as we have seen, originate from the imaginal disks, which remained quiescent during the entire larval stage of development. But in many insects, such as *Drosophila*, the internal restructurings that take place during the pupal phase are not limited to the realization of those organs that were not present in the larva and that will instead be present in the adult. In this case the restructuring also involves the epidermis and the musculature, which originate from an elevated number of histoblasts: cells gathered in small groups distributed throughout the body that survive the systematic destruction of the larval structures that can be observed during the pupal phase. The limited resources available in the pupa are then distributed between all the proliferative centers that will form the adult insect—the imaginal disks and the histoblasts. One should therefore expect an intense competition for resources between these different centers, and this is easy to demonstrate—by suppressing the imaginal disk that would have given rise to a leg or a wing, for instance. In the absence of the imaginal disk, the corresponding appendage is naturally not developed and the resources that have thus been saved are used by the closest imaginal disks, giving rise to structures that are slightly larger than normal.

This competition, moreover, is perceptible only for short distances and is more intense within a single body part than between body parts. This last circumstance perhaps allows us to explain an apparently bizarre aspect of the difference in the horns present (in males especially) on the head and thorax of many scarabs.

In the species belonging to an extremely vast genus of dung beetles called *Onthophagus*, which includes more than 1,500 different species, the horns that decorate the first segment of the thorax are generally much more developed than those present on the head, which may even be absent. In some species, however, the opposite is true: in other words the horns on the head are showier than those on the thorax. This difference might seem like a simple detail, useful only to the species in question (like SMRS) and to entomologists (as a useful criterion for the distinction of species), but it is probable that it will instead reveal a precise connection between a growth problem (which concerns developmental biology) and a problem of sexual selection (which concerns evolutionary biology).

It just so happens in fact that in species in which cephalic horns are more developed the eyes are particularly small. This circumstance leads one to think that, during the pupal phase, the reduced activity of those imaginal disks that lead to the formation of the eyes will make resources available that are utilized by groups of histoblasts from the same part of the body (the head) to build a larger horn. This story might have begun because of sexual selection based on bigger horns. But this would have entailed fewer resources available for other structures of the head, eyes included. The resulting reduction in the size of the eyes would have been "accepted" only at the price of modifying the species' habits, in other words its moving to an environment where good eyesight is less important than was the case with the requirements of the original environment. Or, on the contrary, these beetles could have manifested a tendency to a reduction in eye size comparable to that in cave animals, which is however much more pronounced and has often led to the complete disappearance of these organs. The resources that then became available would have been utilized for the rudiments of horns whose increasing development would have then been stabilized, or even reinforced, by sexual selection. Only an adequate reconstruction of the evolutionary history of this group will aid us in understanding which of these two scenarios is closer to reality. One could additionally suggest a third explanation that uses Amotz Zahavi's so-called handicap principle. The key to this evolutionary choice would then lie in sexual choice on the part of females. Males with long horns have small eyes; but only strong males can allow themselves to see less well, therefore they are preferable to others as partners. But to recognize them it is not necessary to measure the diameter of their eyes, given the correlation between the size of the eyes and the length of their (more visible) horns.

Competition and Cooperation: Two Sides of the Same Coin

Little scarabs, on the other hand, need both eyes and horns. And mandibles, antennae, wings, legs, and so forth. Each of these organs has a cost, not only in terms of exercise, but also, and above all, in terms of investment: the costs, in other words, that the animal must face in order to construct that specific organ. Spending more in the realization of an organ can entail a reduction in the resources available to the other parts of the body.

As we saw, this competition is particularly dramatic in the case of a closed system like the pupa of an insect, but it can be significant even in less extreme situations. In various families of lizards and salamanders, for instance, one can observe that individuals with longer bodies tend to have small or very small, short legs, and with progressively fewer fingers, until we reach the condition of snakes, in which the body is extremely long and the legs have completely disappeared.

An explanation of the correlation between lengthening of the body and size of the limbs in terms of adaptation could invoke the "internal consistency," from the point of view of the mechanics of locomotion, which the opposite modifications of body and limbs lead to. But such an explanation, which is perhaps more reasonable when applied to extreme forms of elongation of the body or of reduction (indeed, disappearance) of the limbs, only with difficulty explains the preservation of the numerous intermediate forms and, above all, the generality of this correlation between trunk and appendages.

In these circumstances also, as in the many other cases we encountered in preceding chapters, it instead seems appropriate to look to developmental biology to understand the phenomena observed. It is possible that, in this case, as with the pupae of scarabs, we are looking at an issue of competition, here between the material available for the growth of a trunk on a path to becoming extremely long, and the material available for the growth of four limbs, which will become increasingly short, if not actually disappear.

Everything Small

The competition between the different parts of the body during development is, therefore, a generalized phenomenon that can have consequences of great importance for the evolution of the form of the adult animal. As is true with many natural phenomena, this competition is more noticeable in extreme situations, and in this case, the best example is probably represented by those evolutionary lines that seem to have taken the path of miniaturization.

A miniaturized animal is not simply a small animal. It is rather an animal that is especially small when compared to its closest relatives.

A tick, for example, is an animal that is much smaller than the victims whose blood it sucks, but no zoologist would ever call it a miniaturized animal. On the contrary, ticks, whose body is normally several millimeters long but can often be well over a centimeter in length after having gorged on blood, are real giants within the mites, the arachnid group to which they belong, where the usual size is more on the order of several tenths of a millimeter. There are also miniaturized mites, like the eriophyids who produce galls on the leaves of plants, but in this case the overall length can diminish to a twentieth of a millimeter, which is smaller than a large number of unicellular organisms!

The diminutive fish, described several years ago with the name *Schindleria brevipinguis*, whose adult females don't exceed nine millimeters in length, males seven, is miniaturized. Loriciferans, little marine animals about a quarter of a millimeter in length, whose body, with an extraordinary degree of structural complexity, is formed by an extremely elevated number of tiny cells, not less than ten thousand, can also be defined as miniaturized.

One cannot in any case deny that there is a certain degree of subjectivity involved in the use of the term *miniaturized*. What should we say of the mites as a group in fact (i.e., ignoring both the giant forms like ticks, and the diminutive ones like eriophyids)? Does their average size, well beneath a millimeter, not justify the designation miniaturized, when compared to those of other arachnids, like scorpions or spiders? But despite the size discrepancy, in this case, it is difficult to precisely reconstruct the events tied to miniaturization, since the mites are an extremely ancient lineage, about whose origins we still know very little. In other words we do not have available terms of comparison, as we do in the case of animals who have recently undergone miniaturization, allowing us a comparison with nonminiaturized species that are related to them. The diminutive fish we just mentioned, for example, belongs to the same family as gobies (Gobiidae), and we can therefore compare it with some of the many not excessively small species of this group.

It is improbable that a miniaturized animal will appear to us as a simple photographic reduction, perfectly proportioned, of its larger relatives. Some organs, in fact, may be arbitrarily reduced, without giving rise to any special problems; others may even disappear, as is the case with respiratory organs (whether they be lungs, gills, or tracheae) that are no longer necessary once we descend below a certain size because the gaseous exchange with the external environment can take place directly, with sufficient intensity, through the body wall. But there are organs that resist miniaturization more than others and this can be related to two different causes: often, it will depend on a little of both.

One of these explanations is of a functional nature: below a certain size, or a certain minimal level of structural complexity, it is possible that an organ may no longer function adequately. An excessive reduction of the brain, for instance, could lead to the disappearance of some centers that the animal cannot do without. The whole would disintegrate, the cooperation between the surviving components no longer sufficient to keep the system alive.

Another explanation refers instead to the opposite pole of our dichotomy: that is to competition rather than cooperation. And it is, naturally, an explanation in terms of developmental biology, which refers to the differing degrees of efficiency with which the first rudiment of the first organs are able to grab a greater overall share of the resources available within the embryo for their benefit. They may succeed simply because they started growing before the other rudiments, in other words because of the more elevated rate at which its cell proliferation proceeded. It is natural that ultimately only those miniaturized animals will survive in which an organism constituted by a sufficiently articulated set of parts capable of functional mutual interaction has been developing. But it is as true that this result, even before being acceptable on the adaptive level, has to be possible on the level of ontogenetic mechanisms. Once again, evolution and development proceed together indissolubly.

Development in Stages

The double bond that exists between different parts of the body, in terms of structures that are developing while competing with one another for the same set of available resources, and as elements of a complex system on whose functional coordination the survival of the whole depends, therefore poses considerable limits on the evolutionary flexibility of organisms. But these are not absolute bonds. There is in fact the possibility of staggering the realization of different structures or the performance of different "tasks" over time—"tasks" that will in any case be completed in the course of a biological cycle. For simplicity's sake we will limit ourselves to considering those animals that reproduce exclusively by means of eggs and spermatozoa; for those that make use of budding or other forms of asexual reproduction there are, in fact, other opportunities (and other problems that we shall not deal with in these pages).

In order to reproduce, an animal must fulfill two conditions. On the one hand it must produce a sufficient number of gametes, and this alone can be a costly operation, above all in the case of eggs, each of which receives a supply, sometimes enormous, of nutritional material. On the other hand the animal with its fill of mature gametes must be able to place them correctly in the environment. For many marine animals this task is simple: it is sufficient to release the gametes in the water, they will worry about meeting. In many other cases, however, fertilization presupposes specific and complex relations between the individuals who produce gametes, especially when fertilization is internal, as is the case with most terrestrial animals, many freshwater species, and also some marine animals. And this is not to mention the diverse and often sophisticated strategies many animals employ to avoid or reduce the competition with other individuals for a partner, or the complex and costly set of physiological and structural adaptations that parental care may require. Finally we should add that for many species reproduction is not exhausted by a unique generative event and the release of a mass of gametes, but goes on, often in a discontinuous manner, for long time periods during which, in addition to the production of new gametes, the animal may have many other "personal" needs to satisfy.

In any case, precisely because their production is extremely costly, an animal cannot engage too soon in the production of gametes. The resources devoted to them would be subtracted from the in any case limited available resources that the somatic structures can draw on during development. Consequently the life of an animal is often clearly articulated in two phases. In the first, the available resources are almost entirely used for growth and somatic development. This is true for the resources that come from the egg and for those possibly received from the mother (in viviparous animals), and for those that the animal subsequently procures for itself during a more or less lengthy period of independent life. In the second phase, a noticeable quantity of resources is instead directed toward the germ line, thus allowing the formation of gametes. An appropriate hormonal control therefore articulates the individual's life into two distinct phases, avoiding a dangerous and useless internal competition. Dangerous and useless because the precocious formation of gametes that would otherwise result would be rendered useless by the inadequacy of a somatic structure that is incapable of placing them in the right place at the right time.

It is not difficult to find a good example of the strong competition that can take place in nature between somatic and germ line, and, therefore, of the importance that a correct chronological positioning of those critical choices, by means of which resources are allocated to one line rather than to the other, has for the animal. In queen ants, the flight muscles are necessary during the swarming from the maternal nest, but their usefulness ends once the nuptial flight, during which the queen is impregnated, is concluded. Once back on the ground, she not only loses her wings, but the flight muscles undergo a true process of degeneration. In actuality these constitute a resource that can be better mobilized to help with the maturation of the eggs. In this case we are going well beyond the simple choice of devoting resources either to somatic growth or to the maturation of gametes. In the case of the queen ant, the same resource is utilized, sequentially, first to the soma's advantage, then to the germ line's, by means of a true internal conversion.

Chapter 11

Making and Remaking

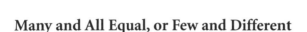

Many and All Equal, or Few and Different

Its recent resounding successes notwithstanding, biology continues to be treated by many adepts of the physical sciences and some philosophers of science as an inferior discipline. Biology, the commentary often goes, has not been sufficiently mathematicized, is not sufficiently predictive. And above all: in biology, laws comparable to those of Galileian science are nonexistent. Many biologists, feeling the sting, respond by calling some of their generalizations—whose range is often modest and, upon careful scrutiny, reveals a validity that is limited to a fairly narrow field—laws of nature.

One of these is known as Williston's law and attributed to the American Samuel Wendell Williston. According to some authors this law would regulate an aspect of biological evolution that concerns structures formed by the repetition of a series of similar parts, like a flower's sepals and petals, or the segments of the body of annelids, or the vertebrae in vertebrates. According to this "law," in the course of evolution these structures would tend to progressively diminish in number and simultaneously to increase in diversity. One would, for instance, pass from primitive flowers endowed with a large and variable number of elements that resemble one another (i.e., not yet differentiated into sepals and petals) to flowers constituted by a small and invariable number of elements, with a clear distinction between sepals and petals. Continuing along this path one would finally reach those families where each of the few elements that forms the exterior of the flower has characteristics that differ from the others, as occurs in many legumes and orchids.

Similarly in the context of annelids, the condition of earthworms, with a generally large and variable number of segments, would be more primitive than that of leeches in whom the number of segments is relatively low (thirty-two as we saw previously) and constant for the entire group.

This evolutionary tendency does seem widespread, but we should carefully investigate how extended and generalizable its reach actually is. It is difficult to argue, for instance, that snakes should be considered particularly primitive reptiles, even if among the diminutive and poisonous blind snakes there are species that can have more than three hundred vertebrae. The same can be said for caecilians, amphibians without legs and a wormlike body that live underground and can reach or exceed two hundred vertebrae. The record for vertebrates with the highest number of vertebrae, however, belongs to some fish that are related to eels: their skeletal axis may have over 750 elements.

Moving on to arthropods, another group of animals whose bodies are clearly formed by a sometimes long series of repetitive elements, our attention has to be directed to myriapods. Among centipedes (the group whose singular predilection for uneven numbers of pairs of legs we have already examined in detail), the record belongs to the exotic *Gonibregmatus plurimipes*, with 191 pairs of legs, which only barely beats a southern European species, the robust *Himantarium gabrielis*, which can reach 189 pairs. These already noteworthy numbers are exceeded by some millipedes that can, so to speak, benefit from the fact that each ring of their body is equipped with two pairs of legs, rather than only one. Not even the longest of millipedes, however, reaches the number (one thousand) that gives them their common name. The millipede with the most legs has to make do with 375 pairs.

There is however another important group of segmented animals, the annelids, and this is where we find some species whose extremely long body comprises more than one thousand segments: a goal that they have exceeded at least twice, with some species actually reaching fifteen hundred units.

We are undoubtedly talking about remarkable numbers, but this *Guinness Book of Records*–type of entry doesn't tell us much, in itself, about the evolutionary tendencies that characterize these groups of segmented animals. What we would like to know is whether the forms with the greatest numbers of vertebrae or segments are really the most primitive within their respective groups, as Williston's "law" would have it. The answer to this question is, in all cases, negative. Let us look at one example.

In the context of centipedes, the relationships between the different groups have been reconstructed in recent years on the basis of anatomical characteristics, as well as of information derived from the comparative analysis of genes and proteins. And everything seems to indicate that it is precisely the condition of those groups of centipedes that are equipped with the smallest number of segments (the scutigeromorphs and the lithobiomorphs, of which the house centipede and the brown centipede, respectively, are common representatives), in other words those with fifteen pairs of legs, that is the most primitive. From this level one passes to forms with a slightly higher number of segments, in other words to scolopendras, which—as we said in chapter 3—are equipped with either 21 or 23 pairs of legs, to end up with geophilomorphs, where this number is increased, going from a minimum of 27 pairs to a maximum, just mentioned, of 191.

But it is not just a question of numbers. In addition to being the centipedes with the lowest number of segments, scutigeromorphs and lithobiomorphs are also those in which the trunk exhibits a more complex structure, while the wormlike geophilomorphs are the centipedes with the most homogeneous trunk, whose segments are very similar to one another. In this zoological class, therefore, evolution seems to have followed a path that is the opposite of what Williston's so-called law would have predicted.

And this is not an isolated case. In millipedes for example, the same story is told. In this case as well, a reconstruction of the probable characteristics of an ancestor common to all of today's millipedes shows us an animal with a modest number of segments and legs. The longest forms, with one hundred legs and more, appear, in this group also, to be among the most recent and innovative.

A Product of Factors

In both cases, that of the centipedes and millipedes, the increase in the number of segments observable in the less primitive families could depend on the discovery, in relatively recent phases of the evolutionary history of these classes of arthropods, of a new mechanism—or, at least, of one scarcely used up to this point—to construct segments in large numbers. As said in an earlier chapter, it seems there is more than one way to construct a segmented animal. Or, at least, the techniques employed by vertebrates in the construction of the spinal column seem to differ from those used to divide a *Drosophila* embryo into segments. It is not even certain that all those parts that are serially repeated along the principal bodily axis *of the same animal* (be it vertebrate, *Drosophila,* or earthworm) are formed in exactly the same manner. Serious doubts in this regard are motivated both by old observations on the embryonic and larval development of various animals, and by recent observations on the location of the products of some genes that are undoubtedly involved in the articulation of the body into segments in specific parts of the embryo.

Of the different mechanisms that lead to the division of an animal's body into segments, some may be more ancient, others more recent. Some are more sophisticated, and one is therefore led to believe that they have only appeared once in the course of evolutionary history; other are simpler, more "generic," enough to induce one to suspect that different evolutionary lines discovered them independently.

In this regard the knowledge gathered by embryologists since the second half of the nineteenth century has been enriched by modern studies on the cellular and molecular mechanisms that generate these segmented structures. But there are still many obscure areas. Above all, it is still not clear up to what point we can generalize from the results of studies on a very limited number of model species that have been undertaken so far. In the case of arthropods, for example, the processes that lead to the subdivision of a *Drosophila*'s embryo into segments are known in detail, but the little we know about bees, grasshoppers, and arthropods that are not insects (e.g., spiders, crabs, and centipedes) clearly shows that the processes are not always, strictly, the same. One encounters significant differences between one group of arthropods and another, even at the level of the genes involved in this division of the body into segments. The same gene, for example, may participate in this process in one species (e.g., in a centipede), while in another species (e.g., the fruit fly), it is excluded from participation, having perhaps taken on another and no less important role in controlling another stage of embryonic development.

In any case, to confront the problem we are now interested in (the validity of Williston's "law") it is sufficient to gather some clues that will allow us to formulate a hypothesis about the evolution of segmentation mechanisms, and therefore about the possibility that these mechanisms may, under certain circumstances, reopen existing options, allowing for a consistent and almost instantaneous increase in the number of segments that make up an animal's body.

Leeches offer a good example. Not so much because of their unique process of teloblast proliferation, which you may remember is the proliferation of those embryonic cells that give rise to a long series of offspring cells that align themselves with those produced by other teloblasts to form the material for the construction of each of the animal's thirty-two segments. Our attention should instead focus on those secondary subdivisions of each segment that have received the designation of rings. The subdivision into rings is obvious but it is limited to the body wall; in other words, it is not found in the internal anatomy, where instead the segmental repetition of nervous ganglia, excretory organs, and other structures is evident. This is therefore an incomplete, partial subdivision, as one could expect from a process that, during the entire course of the embryonic development of the leech, most likely only occurs after the construction of the real segments themselves. It is therefore a sort of secondary subdivision whose impact on the total number of repetitive units that can be observed on the surface of the leech's body is in the nature of a factor in multiplication. And it is easy to perform some calculations. The segments that are affected by a complete subdivision into rings are, in fact, about twenty (those closest to the body's extremities being basically excluded, as we said in chapter 3), and each one is divided into an identical number of rings. Five rings per segment, as is the case with the medicinal leech, adds up to about one hundred rings total. In other species, each segment is divided into three rings, which means that for the entire animal the total is about sixty.

But let us try to imagine what would happen should the interval between the two segmentation processes (the one that gives the animal its basic repetitive structure and the one that controls the further subdivision of each primary segment into a certain number of secondary units) be short—in other words, should secondary segmentation also take place early, before a little heap of ancestor cells manages to differentiate into the various organs that belong to each segment. In this case a multiplication would occur of those centers from which the development of the animal's sections, with all their structures, takes place. We would, for instance, obtain a medicinal leech with about one hundred segments, each with its complement of nerve ganglia, excretory organs, and so on.

These "multiple" leeches, the result of a true double segmentation, exist only in my imagination, but it is probable that a mechanism of this type is truly responsible for the bizarre numerical patterns we found in centipedes. In other words, it is possible that in these arthropods (but then why not also in insects, spiders, and crabs) a more or less fixed, primary-segmental organization exists (identical in all cases), onto which a precocious subdivision of the primary iterative units into secondary ones is later superimposed. And this would occur so early as to basically leave no trace. The two phases of the segmentation process would be completed before the construction of the internal organs was set to begin. In other words, both would affect that phase of development in which the worksites are installed, each of which would then operate independently.

But clues exist that favor the double segmentation hypothesis. There are even cases in which it is difficult to tell whether the animal is comprised of N segments or twice that number. As I said earlier, in millipedes the trunk is subdivided into repetitive units, each of which is equipped with two pairs of legs. This is an unusual situation when compared to that of all other living arthropods where each segment, if it is equipped with articulated appendages, always has only one pair. This is the case with insects, scorpions, crayfish, and centipedes.

In millipedes, instead, the doubt can persist as to whether to count the trunk rings separated by an articulation as segments, or to count each portion that is equipped with a pair of legs as a segment. In the first case we shall resign ourselves to accepting that in these arthropods each segment is generally equipped with two pairs of legs, instead of the usual single pair, while in the second case we shall designate the distinct structural units into which the trunk is articulated as diplosegments (double segments).

In fact this unusual situation could even suggest that our concept of segment is an abstraction, and that nature makes and unmakes these repetitive units—combines them and separates them—according to what the developmental processes and natural selection allow. Construction and adaptation, evo-devo once again.

Returning to the double segmentation hypothesis, it appears evident that nature has given millipedes the possibility of simultaneously constructing both N repetitive parts (the "diplosegments" of the trunk) and 2N parts (the pairs of legs, with some related structures), usually associated with the preceding. Having said this, it is not difficult to conceive that these two numbers, N and 2N, represent two phases in a process of double segmentation, before and after a multiplication event, respectively.

Calculations without Error

The total absence of individual variation in the number of segments that can be observed in many millipedes and centipedes, even among the rather long ones, also seems to favor the double segmentation hypothesis. How is it possible, for example, that in many species of centipedes all individuals, of both sexes, are equipped with exactly forty-five pairs of legs, not one more and not one less?

It is practically impossible to imagine a counting mechanism that would ensure that these animals' embryos stop at the right time (without ever making a mistake!) once the desired number of segments has been reached by adding one segment at a time. The first person to point out the extreme implausibility of such a scenario was

John Maynard Smith, one of the most notable personalities in the field of biology during the second half of the last century. It would in fact be quite difficult for there not to be some degree of variation, in other words some difference between one individual and another within the same species, in the number of units that constitute a series composed of several elements, whether these are the vertebrae of a sardine, the segments of an earthworm, or the petals of an anemone. How is it possible, therefore, that some centipedes never make mistakes?

Knowing almost nothing about developmental biology, Maynard Smith (who, after all, started as an engineer and not a biologist) suggested a possible answer. It is probable, he wrote in 1960, that in these animals the embryo first subdivides into a small number of primary segments, and that each of these then subdivides into a fixed number of secondary units. It is without a doubt easy to exercise precise control on such a mechanism, for instance by first realizing five primary segments, each of which is then divided into two parts, maybe repeating this duplication process once or twice more. In this fashion one would obtain, in a reliable manner (without any errors) somewhere between $5 \times 2 = 10$ and $5 \times 2 \times 2 \times 2 = 40$ segments, depending on how many times the secondary segmentation process was replicated.

But there is more. It is not only the spinal column of vertebrates and the trunk of millipedes and leeches that are divided into se ments. Many appendages are also segmented, like the legs and an tennae of insects and those of millipedes and centipedes themselves. We can therefore ask whether the double segmentation mechanism we hypothesized for the trunk of arthropods is also valid for their appendages. Some clues seem to indicate yes.

One comes from the antennae of beetles, which in the great majority of cases are constituted by eleven articles. This number does not vary from one family to the other, notwithstanding the great differences in the overall appearance of the antenna and its overall length (both in absolute terms and relative to the insect's overall size; fig. 17). Cases in which the number is lower are completely unknown

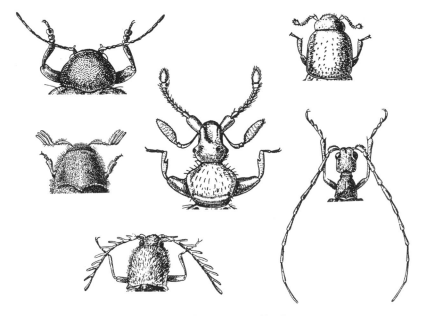

Figure 17. A sample of the diversity of beetle antennae.

in many families, but frequent in others. For instance, in an order as vast as the Coleoptera (at least 400,000 species described, more than a quarter of all known animal species), the cases in which one finds more than eleven antennal articles are extremely few, and in most cases the exception is twelve, even though in a couple of cases the number goes as high as about thirty.

For our discussion of Williston's so-called law, it is significant that antennae with fewer than eleven articles are not more complex (i.e., they do not exhibit a greater specialization of the single elements in the series) than those with eleven elements. At the very least, therefore, we can say that in the antennae of beetles a reduction in the number of parts does not go hand in hand with the greater complexity of the appendage, and vice versa.

Unfortunately the events that lead to the construction of antennae in adult Coleoptera are not easy to observe. We are in fact dealing with insects with a complete metamorphic cycle, just like flies, bees, and butterflies, and the appendages of the adult do not grow gradually, starting from those that the insect has in its juvenile stages, as in grasshoppers, but are formed all at once, during the pupal stage, starting from a rudiment in which it is not possible to discern traces of the future architecture. For the time being we should therefore make do with some calculations, based above all on the frequency distribution of beetle antennae with a number of parts other than eleven. Such calculations make a double segmentation mechanism plausible. But in other insects, such as earwigs, crickets, and cockroaches, in which the antennae lengthen gradually, in correspondence with the various molts the animal undergoes in the course of its life, the existence of primary and secondary segments is a given, just as it is in some crustaceans.

Perhaps the subdivision into secondary segments of a section of the body, or of an appendage that already has clearly defined boundaries (since it is a primary segment and the result of a prior segmentation process), is such an easy operation that animals have discovered it many times over. It is also possible that the mechanisms responsible may not always be the same at a molecular level. In the end, to cut a piece of wood into two parts we can use different tools, like a saw or a hatchet, just as we can use many different types of adhesive to glue them together. What matters is separating in one case, and joining in the other: these are such elementary operations that we can use simple and nonspecific tools to accomplish them.

If this logic can be applied, as I believe it can, to processes of secondary segmentation like those that divide a leech's segments into rings, or those we hypothesized in the subdivision of a centipede's trunk, then we must believe that nature has a simple mechanism at its disposal to go in the opposite direction from that implied by Williston's "law." Several times, in cases that are mutually independent, therefore, a new mechanism capable of realizing a large number of secondary segments would have been added to an older mechanism capable of producing segments (one that characterizes arthropods, and—a rather different one—that characterizes annelids). Not very costly, probably, both in terms of developmental mechanisms and in terms of the corresponding control by genes, and possibly also not very risky in adaptive terms, and perhaps even opening a door to new solutions.

It therefore appears clear that the main limitation of Williston's so-called law lies in its considering the repetitive elements of a certain class (be they petals, vertebrae, or segments) as strictly homologous elements, whatever their number and condition. But this implied premise cannot be taken for granted. In addition, recent progress in the comparative method, aided by research in developmental genetics, invites us to give up the traditional notion of an all-or-nothing homology. As I said in chapter 8, two structures being compared can be considered homologous following a number of different criteria, but not homologous according to others, and it therefore seems appropriate to adopt a criterion for homology that we might call factorial.

Everything we have discussed in these last pages, therefore, seems sufficient to reject the validity of this macroevolutionary principle. But the story does not end here. The fact is that Williston, author of some important monographs on fossil reptiles and of valid entomological studies, never dreamt of enunciating the principle that today bears his name. In actuality, in a work that dates to 1914, Williston simply said that in the skull of reptiles, birds, and mammals no "new" bone had ever appeared, whereas many bones present in so-called inferior vertebrates seemed to have been lost along the way. But Williston himself seemed to be perfectly aware that what is true for the cranial bones is not necessarily true for the vertebrae. Unfortunately, in more recent years, the principle that Williston had intentionally applied to some zoological groups and to some organs, but not to others, was unduly generalized, mostly by American authors. This is how "Williston's law," which is often quoted today in discussions about macroevolution, was born. Additionally, the principle that today goes by Williston's name had already been stated, as early as the first half of the nineteenth century, by several German authors such as Treviranus, Meckel, and von Baer. The works of all these antedate the work by Bronn (1858), to whom the principle was, again incorrectly, ascribed for a while (before his name gave way to Williston's).

Broken Chains

The repeated invention of double segmentation shows how difficult it is to state that some aspect of the body organization of animals has gone down a blind alley and can no longer evolve if not by means of modest variations on a well-known theme or, instead, by taking the path of structural regression, as can often be observed in parasitic animals (for example in the tapeworm, whose ancestors lost their mouths and their entire digestive tract) or cave animals (in whom the eyes almost always disappear).

It seems difficult, however, to overcome certain limits. A wormlike animal, without lateral appendages, can enrich the repertory of its accomplishments if it succeeds in subdividing its longitudinal axis into a series of regions that are specialized by function. For instance at the front end, it can gather the mouth and a certain number of sensory organs, in addition to the brain to which these organs send their information. A subsequent area can specialize in the treatment of food, yet another host the reproductive organs or genital apertures. But all this has a limit, or, rather, two limits.

The most obvious is of a functional nature. An all-purpose tool can get us out of a bind in many different circumstances, but in none of them will it match up to a good, specific tool. The second limit, which is less obvious, but no less binding, concerns the possibility of constructing a wormlike animal along which a number of areas with different specializations follow one another.

In their anatomical descriptions, zoologists often give names to recognizable regions along the animal's longitudinal axis. Head, thorax, and abdomen in insects, for example; prosoma and opisthosoma in spiders. The number of these regions generally varies between two and four. These are, naturally, arbitrary descriptive outlines, but it is still significant that zoologists don't know of any animals along whose body ten or twenty such regions exist. Here we are no longer talking about simple repetitive units like segments, whose number can be high, but about regions with different specializations. In segmented animals, each region may include a certain number of segments; in insects, for instance, one usually observes six segments in the head, three in the thorax, and eleven in the abdomen.

It seems that developmental mechanisms are incapable of producing longitudinal axes of significantly greater complexity than what is observable in known animals. From this point of view, then, there seems to be a limit in the possibility of evolving toward more complex forms. But there is an alternate possibility, and some zoological groups seem to have discovered it. If not much more can be accomplished along the body's main axis, would it not be possible to work *on multiple axes* in the same animal? Translated into the language of anatomy, this question is equivalent to asking oneself how and if it is possible to construct appendages. Something familiar, if we think about our four limbs. Something that is common to two particularly successful zoological groups: arthropods and vertebrates. Something, however, that is far from being universal and, quite probably, was not present in the earliest animals. A body's appendages constitute an evolutionary novelty that appeared independently several times.

Certainly not all appendages possess the same structural complexity or importance for the animal. The tentacles that surround the mouth of a hydra, or of a coral polyp, for instance, are very simple, and the same can be said of the tube feet by means of which a starfish attaches itself to a reef and which allow for its slow movements. Things change if we shift our attention to a butterfly's antennae, the legs of a grasshopper, or the limbs of a hare. Certainly no one could expect appendages as complicated as an insect's antennae in an animal as simple as a hydra, and maybe, even if with less conviction, no one could expect exceedingly simple appendages in a complex animal like a vertebrate or an insect. Certainly but why? For what reason should the main body axis of an animal, and the secondary axes that the appendages represent, as a matter of principle, exhibit a comparable level of complexity?

From the current perspective of developmental biology, where genes are masters, one might answer this question by saying that a hydra's genome is only capable of realizing a simple structure, such as the polyp's body, which is like a little sack, or the tentacles that surround its mouth; vertebrates instead have a more sophisticated genome, capable of producing much more complex animals, both in their main axis and in their appendages. The reason is not clear, however, why a condition that one presumes is sufficient to generate complexity (let us say the genome of a vertebrate, or an arthropod) should then be transformed into a handicap that impedes the realization of simple appendages in animals with a complex trunk.

In the last fifteen years this scenario has been enriched by an important discovery in the area of developmental genetics. In complex animals such as vertebrates, a certain number of genes, belonging to the group of those that are important in giving structure to the animal's main axis, are also involved in the specification of parts along each appendage's axis.

Co-option

The dual involvement of these genes is usually "explained" in terms of co-option. Briefly, the interpretation proposed is the following.

There was once an animal with no appendages, but with a trunk articulated into regions, constructed under the control of numerous genes, among which are the *Hox* genes that we have mentioned. One of its descendents, however, learned how to build appendages. Initially these were simple, and not too highly structured, but at a certain point, some genes that had already been involved in the specification of particular structures along the animal's main axis were now also expressed along the axes of the appendages and were co-opted here to undertake functions new to them. The final result is the production, in complex animals, of those complex appendages that they today possess.

The proposed scenario might be attractive because of the gradualism of the transitions it suggests. However, I do not regard it as very plausible. My objection does not concern co-option in and of itself, an undoubtedly reasonable mechanism that in other cases very probably does represent a path that led to significant innovations in the course of the evolution of animal body plans. The problem instead lies in the fact that in the story we just told nothing is said about the most important event: the origin of the appendages. One is concerned with an explanation of the current involvement of genes—which originally only had a role in giving the animal's longitudinal axis a structure—with the appendages, but the presence of the appendages themselves remains without explanation. In all of this, it seems implicit that building appendages is something quite different from building the body's longitudinal axis. The appendages' story would therefore start again from scratch, from a minimal structure (but one attached to an already complex body!), which in the course of evolution becomes increasingly complex by means of the progressive co-option of genes that, little by little, in addition to being expressed in the trunk as they had been for a long time, are now also expressed along these new secondary axes.

This scenario is not very plausible and also has gaps. It has gaps because it does not explain the origin of the appendages, limiting itself to a story about how these became progressively more complex; it is not very plausible because it leads us to believe that the first appendages must have in any case been formed independently from the animal's trunk and in a different manner.

But how is it possible to hypothesize two distinct mechanisms, both capable of building a body axis in the same animal? Two mechanisms, moreover, that are able to operate without interfering with one another, and yet so similar as to allow one of them, sooner or later, to co-opt quite a number of components from the other? And, what is more, why would all these co-options only go one way, from the trunk to the appendages, and not vice versa?

Buds and Appendages

In my opinion all these difficulties can be overcome if we are willing to take a little conceptual leap, if we are willing, in brief, to consider the appendages as a sort of copy of the body's main axis. Copies, naturally, that have been revised and corrected, but copies, nevertheless, whose origin cannot be explained by ground-breaking events, at least from the point of view of developmental biology.

That an appendage might be a sort of revised and corrected bud might seem a doubtful proposition if not worse. But let us try to reason a little. Nobody takes exception to the body of a polyp, a hydra for example, which can form a bud that is destined to become a complete copy of its parent. Some might be led to concede that the hydra does exhibit this kind of behavior, more like a plant than an animal, precisely because its structure is among the simplest in the entire animal kingdom. But the ability to reproduce by means of lateral buds also exists in some polychaete annelids whose structural complexity does not differ much from that of an arthropod; some tunicates can also reproduce by means of buds, and they belong to the same phylum, the Chordata, as vertebrates.

Certainly starting a bud is not a banal operation because it requires the activation, at a precise point on the parent's body, of a series of cellular dynamics that differ from those that would normally occur along the trunk and would instead compete with them. But it would be wrong to think that these dynamics are very different from those that lead to the realization of a new individual throughout the usual stages of embryonic development. There will certainly be some differences, but they are probably limited to the initial, triggering phase. And actually, in colonial sea squirts, in which the production of new individuals by means of buds and the generation of offspring by means of fertilized eggs coexist (or, rather, alternate), from a certain developmental phase on, the differences between individuals born in these different ways become negligible.

In other words, in order to build a new body axis the animal must resolve some internal problems, but it does not need to invent an entire system of new mechanisms controlled by an unknown number of genes. The genes and the mechanisms to build an axis are already at its disposal and are none other than those genes and those mechanisms that come into play to build the animal's main body axis.

Certainly, to transform a bud capable of reproducing the body axis of an animal without appendages into an appendage of the animal itself, requires traveling down a relatively long path. But the scenario becomes less problematic if we remember that this history, if it really occurred in this fashion, started a long time ago, when the main body axis of the animal capable of producing buds would have differed from that of a vertebrate or an arthropod, and was perhaps simpler. From that time on, not only would the bud/appendages have evolved, but so would the trunk that generated them.

A decisive event, along the path of the possible transformation of a bud into an appendage, could have been the exclusion from the bud of material deriving from the internal leaflet, in other words the absence of a lateral branch of the digestive tract. But perhaps it is best not to proceed further in an excessively hypothetical reconstruction. What I instead want to underscore is the parsimonious nature of a hypothesis like the one just proposed, which is in harmony with one of nature's general principles: using and re-using what is available, gradually introducing modulations whose effects, in the long term, can be resounding.

The appearance of appendages represented an evolutionary novelty of enormous importance, and not only on the plane, which is perhaps the most obvious, of locomotion. Let us look at arthropods, for example. Here we find the antennae of insects, capable of exploring a world of signals, chemical above all, which a series of receptors localized on the body's surface would never be able to distinguish so precisely. We find the oral appendages of insects: masticatory in grasshoppers, stinging in mosquitoes, lapping in flies; and in butterflies they are transformed into a straw used to suck nectar from flowers. Or, again, the spider's poisonous appendages, or those, armed with chelae, of the crab. And we could continue with the raptorial legs of the praying mantis, ready to be suddenly projected against their prey, or with the thin ovipositor by means of which many wasps introduce their eggs into the tissues of a caterpillar or another victim.

These appendages have traveled a long way, starting from the day they appeared on an ancient segmented animal whose traces have been lost. One can certainly say that the extraordinary success of the arthropods depends, to a considerable extent, precisely on the presence and versatility of their appendages. A success that appendages have also contributed to, if perhaps to a slightly lesser degree, in the case of polychaetes, and that is fully evident in vertebrates as well, notwithstanding the fact that snakes (and with them the caecilians, worm lizards, and several true lizards, like the European glass-snake) demonstrate that one can also live without.

Chapter 12

Innovations without Plans

Regularity without a Program

Not every city has an urban plan with the geometrical simplicity and purity found in Palmanova's nine-pointed star or which reveals the dreary regularity of many American centers planned around a double system of parallel axes oriented toward geographic parallels or meridians. Many cities, especially among the most ancient ones, never had a city plan. They grew according to small local rules, often rising from multiple nuclei conditioned by the geometry of the locales and by the irregularity of the terrain, nuclei that sooner or later merged into an urban aggregate where, the absence of an overall plan notwithstanding, an integrated structure eventually emerged, with a logic and functionality of its own. In the forms of living organisms, there are also regularities that are not the result of a plan, but can be traced back to the simple geometry of the relations among its component parts.

Inexpensive Symmetry

Most animals, for instance, have a body organized according to bilateral symmetry. Any departure from this basic condition appears as a disfigurement or a handicap. This is the condition of the lame, and also of those who have one eye that sees less well than the other.

It is so important to possess a pair of symmetrical eyes and pairs of arms and legs that are mirror-images of one another that we might want to believe in the existence of genes capable of exercising a particularly strict control on this symmetry. But we would search in vain.

Genes for symmetry probably do not exist. Neither are there specific genes for the left hand or the right eye. Genes only exist whose products influence the shape of the hands and the eyes. In addition, other genes contribute to limit the number and position of those places where an embryo may construct a hand or an eye. To make use once more of an image used in a preceding chapter, these define the number and position of the worksites where a copy of these organs will be formed.

At this point the game is over. The realization of a hand or an eye involves a large number of genes, but at each of our worksites, in other words in the groups of cells from which these organs take shape, all the genetic information necessary is present.

Genetic information is not everything, however. The position, relative to the rest of the body of the morphogenetic center from which the eye or the hand are taking shape, is also important. It will be this position, and not a specific gene, that will dictate the polarity of any single hand or eye. The right hand is the only type of hand that the worksite on the right is capable of producing. The organs' regular symmetry is only an amplification, a projection, of the embryo's elementary geometry. The problem, in the history of development, is shifted backward, until one finds that moment in which the embryo acquires its polarity once and for all. Once the front and the back, the above and below have been fixed, right and left are also unambiguously specified.

In some animals these fundamental coordinates are fixed very early. In many eggs, one can recognize a distinct polarity even before fertilization. For example, because of the fact that at one end of the gamete we can find some specific molecules (the so-called germ plasma) that after the fertilized egg's first divisions will be inherited by a small number of cells, the ancestors of the germ line, they will be located at the embryo's rear end. In other cases the uncertainty as to the localization of the future body axes is resolved at fertilization: the animal's symmetry plan is then defined by the position of the sperm's point of entry, in others still by the unequal distribution of the yolk within the egg. This initial stamp will dictate the bilateral symmetry of the animal's entire architecture. No symmetry gene was necessary to give the egg, or the early embryo, its regular geometry. And no symmetry gene will be necessary in the subsequent phases, when the animal's structure will be completed with all its organs.

This argument can naturally be made also in regard to another widespread regularity: the repetition of segments along the body's main axis. In this case as well, we will not have to search for the genes responsible for the shape of our fourth thoracic vertebra, or for the seventy-fourth segment of an earthworm. There will simply be genes involved in the making of vertebrae or segments, and genes that, with greater or lesser precision, will determine their number.

If we want to find genes with a more limited sphere of activity, we must search for those that give a vertebra its identity as thoracic, rather than cervical or lumbar. And in the case of the earthworm, it will be legitimate to find out what the genetic control on the formation of those slightly fuller, and differently colored, segments is, those that can be easily observed about one quarter of the way down the animal's body and that zoologists have designated as the earthworm's clitellus. Otherwise one vertebra is as good as another, one segments is as good as the next. They are ultimately simply replicas and it is not that important if they are not perfectly identical to one another: they are "handmade" copies, made without the use of a strictly cast die.

The world of the living, therefore, is filled with inexpensive regularities. And almost perfect ones. We are animals that exhibit bilateral symmetry, even if the fingerprints of our right hand differ from those of our left. We should, instead, keep our eyes peeled when the repetition of parts is too precise to be the result of free-wheeling replicas. And this is the reason we previously dwelled on the extraordinary regularity in the number of segments of many millipedes. A regularity that cannot be the fruit of chance, but makes one presume a sequence of events that would make error unlikely.

Breaks in Symmetry

If we consider the situation carefully, however, those forms that are not symmetrical can be among those that are most difficult to build. I am not referring, naturally, to those small and accidental differences between the two sides of the body that can be observed, as I just mentioned, in the small reliefs on the skin of our fingers or in the layout of the veins of both arms. I am instead referring to those asymmetries, typical of the species, that can affect the entire body, as in snails and hermit crabs, or only inner organs, like our digestive tract, our lungs, and our heart.

If it is true that a symmetrical structure does not require particular instructions for its realization, a precise departure from symmetry must necessarily be the result of a precise departure from an otherwise symmetrical development, and it is legitimate to suppose that there may be genes involved in this situation. In actual fact we know that in the case of vertebrates, some genes are involved in breaking the animal's symmetry plan, which would otherwise basically be bilateral. These genes act very early in the embryo, before the individual organs, such as the heart, the lungs, and the stomach—which will ultimately by asymmetrical in terms of shape and position—begin to grow.

In these matters of symmetry, as well as in many others we discussed in these pages, development has the first word. Natural selection, if and when it is involved, enters the scene at a later time. It is probable that the latter has (and has had) only a secondary role in preserving the bilateral symmetry shared by most animals. Little departures from it are probably irrelevant to an individual's success, whereas it is likely that more significant (inheritable) departures will accompany other defects that can be more important to an individual's success than asymmetry in itself.

Natural selection does, however, have an important role in supporting the evolution of the asymmetrical body of hermit crabs, since this condition represents an optimal adaptation to a life inside the empty shells of gastropod mollusks.

The origin of the torsion of the visceral mass of hermit crabs, slugs and snails of the earth, freshwater, and sea is still unaccounted for today, after many years of study and many theories. It is in any case quite possible that the adaptive advantages in terms of an appropriate distribution of bodily masses and the appropriate placement of gills, and of anal, excretory, and genital orifices may have been of great importance in the matter.

An interesting aspect of asymmetry is that in most cases only one of the two forms, mirror images of each other, which it would be possible to imagine, actually exists. Almost all human beings, for instance, have a stomach that sits to the left, a liver to the right, and so on. A specular symmetrical arrangement, the so-called *situs viscerum inversus*, is rare. But this does not mean it is pathological. If it were not for medicine, which induces us to check how we are made inside, many cases of *situs viscerum inversus* could easily have passed unnoticed. But, what is more important to our discussion of development and evolution is that the passage from one condition to the next (stomach on the left and liver on the right, to stomach on the right and liver on the left) corresponds to a pointlike genetic difference, the simple shift of the lever of a switch toward one of two symmetrically specular alternatives. This is the reason why the two forms can appear and also coexist in the same species, or, rather, in the same population.

The Origin of Evolutionary Novelties

It is not easy, therefore to foresee the degree of difficulty represented in nature's passing from one form to another. No one would have imagined that it is easy for a scolopendra to pass from twenty-one to twenty-three pairs of legs, or vice versa, whereas a trunk with twenty-two pairs seems really out of reach. Nobody would have imagined that mammals can easily evolve necks of different length by modifying only the shape, but not the number, of their cervical vertebrae.

Evolutionary transitions from one form to the other are a little like the movements of chess pieces. Only by knowing the rules of the game can we understand which squares a knight can reach by moving from its current position, and which squares, instead, can be reached by a bishop or a rook in a certain number of moves. Once again the extraordinary fruitfulness of an interpretation of the biological world that unites issues of development and adaptive evolution comes to the fore.

Since Darwin's times, evolutionary biology has concentrated its attention almost exclusively on problems that can be defined as those of microevolution, where population is the protagonist in the temporal dimension of so-called ecological time—that which can be measured in individual, tens, or, at most, hundreds of generations. It is precisely within this framework that the attention of the biologist is directed to the variations over time of the frequencies of different genic alleles and their assortments. And it is in this area of research that we find experiments attempting to reveal the effects of selection on a population, for instance by favoring larger or smaller individuals, or those with a lighter or darker coloration. Lurking in the background of this research and these interpretative models, however, is a series of questions with a different reach, those concerned with so-called macroevolution. For instance, how was the first bird wing developed? What is the origin of the brain? Where did the first flower come from?

For many researchers, these are false problems. If evolution has progressed by means of the slow accumulation of small differences, generation after generation, the pathway we would like to trace between dinosaurs incapable of flying and those who were able to stay in the air thanks to fore limbs transformed into wings is arbitrary. And the pathway between animals endowed with a brain and their most recent ancestors who still lacked one, or between plants with flowers and their most immediate ancestors who lacked them, would be just as arbitrary.

In other words, evolution would be just microevolution, and the "great events" that so forcefully captivate our attention would be nothing more than macroscopic differences that are perceived when we compare organisms from sufficiently different geological eras along the same evolutionary line or organisms belonging to divergent evolutionary lines. What need is there to hypothesize the existence of a distinct class of phenomena (the macroevolutionary ones precisely) whose comprehension does not seem to require explanations that differ from what we already know at the microevolutionary level?

The issue, however, is not this simple. Evolution is not only a story about the variations in the allelic frequencies of different genes. Natural selection, as we well know, does not "see" genes, but the phenotypes, in other words animals and plants in the full concreteness of their anatomical structure and everyday behaviors. And the relationship between gene and phenotype is neither simple nor unilateral, for a variety of reasons that we have touched upon in the pages of this book.

On the one hand there can be different ways to construct almost identical phenotypes; on the other, the same genotype can sometimes translate into very different phenotypes, depending on the environmental conditions that push development in one direction rather than in another. In many cases, moreover, pointlike, or in any case modest genic differences, can translate into important phenotypical differences, but just as frequently the reverse can occur—that is, changes that affect a non-negligible portion of the genome result in the preservation of the same form. It therefore does not seem fruitful to address the problem of the origin of evolutionary innovations starting from the genes. We should instead recall that each organic form, whether traditional or innovative, is the result of a developmental process. The appearance of new forms must therefore depend on some novelties emerging at the level of development.

Everyone agrees about this. The problem consists in the fact that modern biology has been dominated by the idea of the gene's omnipotence, and this is where the dangerous metaphor of development as the execution of a program inscribed in the genome comes from. It is a short step from here to reducing the origin of evolutionary innovations to a simple game of genic mutations that introduce some changes into an old program. But, as I said, this is a road that does not lead far because it chose the wrong units to describe and interpret the phenomenon.

Modules

If we are searching for a house in which to live, we gather information about the number, the size, and the arrangement of the rooms, and not on the number, the size, and the mutual arrangement of the bricks the house is made of. This does not mean that bricks are not necessary to build a house—nevertheless, we probably won't talk about them in our visit to the real estate agency. And even the draftsman worries about bricks (their quality and price) only after having drawn up the plans for the new house, locating the rooms, the corridors, the stairs, the roof.

In the same way, with regard to development, we can recognize operational units that are more important than genes. In fact we can find more than enough such units—we only have to choose. We can, for instance, consider individual organs, such as the brain, the lungs, and the liver. However we should remember that these organs (and this is all the more true the more complex the organ) are well-defined anatomically and structurally, but not developmentally, as I mentioned when expressing my strong doubts relating to the issue of organogenesis. A more natural choice in the case of vertebrates might be the individual bones of the skeleton, insofar as each one begins its formation at one specific ossification center, or from a small number of such centers. In this case, therefore, an anatomical unit tends to correspond to a unitary process. And this is the path that leads to the concept of a module.

In modern developmental biology, each sequence of events that is able to proceed in a largely autonomous fashion relative to what is occurring around it is called a module. This is a bit like the conversation that takes place between a small group of people in a crowded square, which can basically continue undisturbed for a certain period of time, while the same process is taking place, not far away, in other distinct "conversation modules." Just like these, the modules that can be recognized in the course of development differ in terms of length and robustness, and they can even divide or come together. At least for a while, however, they represent places in which a dynamic process basically proceeds of its own accord. In the case of development, this can also lead to the realization of a distinct anatomical part. Naturally each of these modules involves the expression of a certain number of genes, or, rather, of many genes with more or less significant differences from one module to another. But right now we are not interested in looking *within* each module, or in learning which bricks were used in the construction of a specific wall. We are interested in the rooms, in their reciprocal relations. In other words we are interested in the relationships *between* the different modules.

A Glance at the Clock

Speaking of relationships, it is only natural to address two different dimensions, the temporal and the spatial. And this is precisely where we encounter a terrain in which evolution has been able to indulge its whims and continuously give life to new combinations. If the organism that is developing can be disassembled into modules endowed with a certain reciprocal autonomy, it should then be possible to modify their spatial and temporal relationships. At least this should be possible in terms of developmental "mechanics." Natural selection will later make a decision as to the validity of these new solutions. The first step, in any case, is to try and produce them. In fact a very large number of evolutionary innovations seem to have occurred by taking this path: that is, by modifying the relationships between the different modules.

First of all there are the heterochronies, the changes in the temporal sequence of events. These changes, naturally, do not inevitably lead to momentous consequences, but it is not prudent to guess too self-assuredly.

Let us think for example about the order in which the fingers of a hand are formed. What is the difference between going from the thumb toward the little finger or from the little finger toward the thumb, when ultimately the fingers are still five? The fact is, as we mentioned in a previous chapter, that for some reason, the hand may not ultimately always have five fingers. It could be formed with four, for instance. In this case we should expect that the missing finger would be the little finger in the first case, and the thumb in the second. That is, naturally, once we grant that each finger will preserve its original identity even on a "reduced" hand.

A frequent type of heterochrony is that in which sexual maturity is reached rapidly, and the animal is therefore already able to reproduce even though its overall somatic structure is still of a juvenile or even a larval type. We already encountered one outstanding example in the case of the beetle *Micromalthus debilis* in which the larvae are in charge of reproduction.

Paradoxically the least obvious consequences of heterochrony (but in the long term they are perhaps the most important) derive from the fact that the developmental modules whose temporal sequence can be changed because of the significant degree of autonomy that exists between them do not remain forever isolated from one another. Modules that previously had no way of interfering with one another can now be exposed to new forms of interaction, and completely new processes and structures may result from this state of affairs. The following story can serve as an example.

John and Louise have never met, even though every morning they both stop at the Café Modern at Times Square. They have never met because John has to be in the office by 8:30, and stops for coffee at around 8:15, while the store Louise works in opens at 9:00 and her coffee time is around 8:40. But one day Louise has to be at work half an hour earlier than normal to assist in redecorating the shop window. She enters the Café Modern at around 8:10. Because it's not her "usual" time, she doesn't know anyone in the establishment. Five minutes later, as she is about to leave, she meets John who is entering at 8:15, on time as always. She exits, he asks the bartender where the beautiful girl who just left works. Should John show up, in the following days, in the vicinity of Louise's store, it will be the fault (or merit) of a heterochronous cup of coffee.

At the end of the 1980s, the discovery of defects caused by the failed functioning of a gene (*lin-14*) in the famous *Caenorhabditis elegans* stirred up some interest. In fact in this mutant, some groups of cells that normally appear only two larval stages further on, appear early. At the time people spoke of the first discovery of a gene with a heterochronous effect. And actually a pointlike genetic differ- ence can be sufficient to modify the moment, in the course of devel- opment, at which a specific module starts, or the moment at which it ceases to function, but the consequences of such a change, as we have seen, can snowball.

In genetic terms, a heterotopy can be relatively simple, in other words a change in the spatial arrangement of the different modules. We have already discussed genes like the *Hox*, which control the position of different structures along the body's main axis. Certainly, the *Drosophila* mutants, bred in our laboratories with a pair of legs in place of antennae cannot be taken as a model of evolutionary in- novations, and thus Gehring's fruit flies with an eye segment on their tibia even less. These aberrations, however, do provide us with tangi- ble proof of the potential for dissociation of developmental modules affecting different parts of our body, and they open the horizon to a vista of endless combinations wherein development proposes and selection chooses.

Epilogue

Is everything easy, then? Was it sufficient to elongate seven cervical vertebrae to obtain a giraffe's neck? Was the simple mutation of a *Hox* gene sufficient to give rise to the first insect with only two wings in- stead of four? Does one really need something more than the simple regulation of cellular proliferation to give rise to a seven-millimeter fish or to a thirty-meter whale? It would be a serious problem if a hurried reading of these pages led readers to believe this.

It is certainly true that the mutation of a single gene can have out- standing consequences for the appearance of the animal that carries it, just as it is true that the more rapid growth of one part of the body, compared to others, can result in an imposing change of shape. But a giraffe cannot be reduced to its neck, and its neck cannot be reduced to a series of vertebrae; by the same token, the difference between a housefly and a butterfly does not only consist in the fact that the first has two wings while the second has four; and the difference between the diminutive *Schindleria brevipinguis* and the giant *Balaenoptera physalus* doesn't disappear simply by making a photographic enlargement of the small fish. The vertebrae of the neck, the number of wings, or the overall size of an animal are important as well as showy aspects of its organization, but they are only a tiny fraction of the whole. In addition to the long neck, the giraffe also has a heart, a brain, and four legs; as a larva, the butterfly is a caterpillar with many legs, while the fly is a maggot devoid of appendages; in order to breathe *Schindleria* uses gills while the fin whale uses its lungs.

With this I don't mean to say that in order to forecast the weather in Texas we need to know if at this moment a butterfly is beating its wings in Brazil. In the development of an animal, it remains the case that one can recognize local dynamics that for some time remain basically isolated from everything else that is going on in the embryo. One part of the body, therefore, could be modified, because of a change in the local developmental dynamics, without the rest of the body being greatly affected. One still has to wait and see what natural selection thinks about it, however. In other words it is possible that the modified part will continue to form a functional integrated whole with the rest of the body; but it is also possible that things will turn out differently. It may be a good thing to bring the mouth and brain to a height of almost six meters, but you cannot accomplish this if the heart is not able to pump the blood to that height.

The living organisms that inhabit our planet find themselves, therefore, at the crossroads of two logics, the developmental and the evolutionary. Both need to be satisfied, and we won't travel far if we attempt to interpret the history of biological forms exclusively in terms of genic expression and variations in allelic frequencies.

How could we explain the passage from a gill-based respiratory system to a pulmonary respiratory system in vertebrates, relying exclusively on genes? How could we explain the fact that lungs are not modified gills, but organs with a different origin, which, in a transitional phase, coexist with the gills, allowing for a true amphibian existence? We can perhaps conclude these pages with this important transition that we have benefited from. The lungs in terrestrial vertebrates are not modified gills, but this doesn't mean they come from nowhere. They derive, instead, from a type of dead-end sack that is connected to the digestive tract and widespread in bony fish, where they are mostly not involved in breathing. Basically the novelty consists in their makeover as organs assigned to gaseous exchanges outside of the water.

Nature doesn't have a draftsman that can indulge his whims in exercises of unfettered creation. It must always start from what it has already learned to produce and that, at this point, seems to have proved its worth.

Recommended Readings

—————◆◆◆—————

The first book to address the problem set of evolutionary developmental biology, but without explicitly using this term, is the by now classic but still stimulating volume by R. A. Raff and T. C. Kaufman, *Embryos, Genes and Evolution: The Developmental-Genetic Basis of Evolutionary Change*, New York: Macmillan, 1983 (2nd ed. Bloomington & Indianapolis: Indiana University Press, 1991). The new discipline acquired its identity with the publication of the first edition of B. K. Hall, *Evolutionary Developmental Biology*, London: Chapman & Hall, 1992 and 1998.

Raff and Kaufman are two representatives of the group that has so far dominated evolutionary developmental biology, the one that comes from developmental genetics and found authoritative expression (too technical for beginners) in the volume by E. H. Davidson, *Genomic Regulatory Systems: Development and Evolution*, San Diego: Academic Press, 2001. S. B. Carroll, J. K. Grenier, and S. E. Weatherbee, *From DNA to Diversity: Molecular Genetics and the Evolution of Animal Design*, Malden, MA: Blackwell Science, 2001, 2004, and A. Wilkins, *The Evolution of Developmental Pathways*, Sunderland: Sinauer Associates, 2001 are more accessible and more didactic in their approach. And an even more enjoyable read is provided by the brilliant and nimble volumes by S. B. Carroll, *Endless Forms Most Beautiful: The New Science of Evo Devo and the Making of the Animal Kingdom*, New York: W. W. Norton and Co., 2005, and E. Coen, *The Art of Genes: How Organisms Make Themselves*, Oxford: Oxford University Press, 1999. The latter volume also deserves special mention because it is the only one among those cited in this bibliography that is written by a plant specialist rather an animal specialist.

In the meantime R. Raff has returned to these topics with *The Shape of Life: Genes, Development and the Evolution of Animal Form*, Chicago and London: University of Chicago Press, 1996, in which he offers a summary of the paleontological history of animal life; an even more successful view of the integration of paleontology and developmental genetics is provided by J. Valentine, *On the Origin of Phyla*, Chicago and London: University of Chicago Press, 2004.

More nuanced interpretations of the subject matter, in which developmental genetics does not monopolize the interpretation of the evolution of organic forms, are J. C. Gerhart and M. W. Kirschner's *Cells, Embryos and Evolution*, Malden, MA: Blackwell Science, 1997, and A. Minelli's *The Development of Animal Form: Ontogeny, Morphology and Evolution*, Cambridge: Cambridge University Press, 2003.

For the reader interested in a brief but pithy summary of the modern tendencies of evolutionary developmental biology, a good article is W. Arthur, "The Emerging Conceptual Framework of Evolutionary Developmental Biology," in *Nature*, No. 415 (2002), pp. 757–64. To help with orientation in a literature that is becoming more complex and nuanced every day, the roughly fifty entries in the volume by B. K. Hall and W. M. Olson, *Keywords and Concepts in Evolutionary Developmental Biology*, Cambridge, MA and London: Harvard University Press, 2003, are also very useful.

At the end of the 1990s the first journals dedicated to this new discipline began to appear: *Evolution and Development* and the Molecular and Developmental Evolution section of the *Journal of Experimental Zoology*.

Chapter 1. Unity in Diversity

The old volume by P. Belon du Mans, *L'histoire de la nature des oyseaux, avec leurs descriptions, & naifs portraicts retirez du naturel, escrite en sept livres*, Paris: Guillaume Cavellat, 1555, is once again more easily accessible thanks to the reprint published by Droz, Geneva, 1997.

An accurate historical reconstruction devoted to the figures of Georges Cuvier and Étienne Geoffroy Saint-Hilaire and their disputes is provided by T. A. Appel, *The Cuvier-Geoffroy Debate: French Biology in the Decades before Darwin*, New York and Oxford: Oxford University Press, 1987.

Chapter 2. Archetypes

Currently the most informative source for the history of animal morphology from antiquity to the middle of the nineteenth century is the book by E. S. Russell, *Form and Function: A Contribution to the History of Animal Morphology*, London: Murray, 1916 (reprint Chicago and London: University of Chicago Press, 1982); for the topics we are concerned with here,

Russell's book is useful both in regard to Cuvier and Geoffroy, and for the exhaustive presentation of Owen's morphology.

The work by T. H. Huxley, *The Crayfish: An Introduction to the Study of Zoology*, London: C. K. Paul, 1879, is currently available in a 1974 reprint, Cambridge, MA and London: MIT Press.

CHAPTER 3. EASY NUMBERS, FORBIDDEN NUMBERS

The first important work to gather a lot of data about the variation in the numbers of repetitive parts in the bodies of animals (e.g., teeth, fingers, and vertebrae) is the classic work by W. Bateson, *Materials for the Study of Variation Treated with Especial Regard to Discontinuity in the Origin of Species*, London: Macmillan, 1894, of which there is now also a reprint, Baltimore: Johns Hopkins University Press, 1992.

A possible explanation of the invariance in the number of cervical vertebrae in mammals has been proposed by F. Galis, "Why Do Almost All Mammals Have Seven Cervical Vertebrae? Developmental Constraints, 'Hox' Genes, and Cancer," in *Journal of Experimental Zoology (Molecular and Developmental Evolution)*, No. 285 (1999), pp. 19–26.

CHAPTER 4. PRIVILEGED GENES

A perspective on evolution and development in which genetic information plays a central role is offered, as previously mentioned, in most of the works cited at the beginning of this bibliographical note. The most typical expression of this line of thought, however, is the largely autobiographical essay by W. J. Gehring, *Master Control Genes in Development and Evolution: The Homeobox Story*, New Haven and London: Yale University Press, 1998.

A healthy reaction to this "genecentric" perspective is expressed very convincingly in the excellent article by H. F. Nijhout, "Metaphors and the Roles of Genes in Development," in *Bioessays*, No. 12 (1990), pp. 441–46, and in a more discursive and popular guise in the book by E. F. Keller, *The Century of the Gene*, Cambridge, MA: Harvard University Press, 2000.

An excellent introduction to the science of networks is the recent book by A. L. Barabási, *Linked: the New Science of Networks*, Cambridge, MA: Perseus Pub., 2002.

CHAPTER 5. EVOLUTION AND DEVELOPMENT

The concept of a "space of forms" in which currently existing and extinct organic forms, and those whose existence we can imagine as a result of the combination of appropriate descriptive parameters, find a place, is examined in detail, with examples and applications in G. R. McGhee Jr., *Theoretical Morphology: The Concept and Its Applications*, New York: Columbia University Press, 1999.

Other alternatives to an excessively reductive and univocal perspective on evolution and development with the gene at their center can be found in the volume edited by G. B. Müller and S. A. Newman, *Origination of Organismal Form*, Cambridge, MA and London: MIT Press, 2003. Some of the essays it contains are based on a theory known as Developmental Systems Theory, and for this theory, one should refer to S. Oyama, *The Ontogeny of Information*, Cambridge: Cambridge University Press, 1985 (2nd ed. Durham, NC: Duke University Press, 2000), and to the collective volume by S. Oyama, P. E. Griffiths, and R. D. Gray, *Cycles of Contingency: Developmental Systems and Evolution*, Cambridge, MA and London: MIT Press, 2001. For a critical evaluation of the relationships between the "orthodox" version of evolutionary developmental biology and the view offered by Developmental Systems Theory, the article by J. S. Robert, B. K. Hall, and W. M. Olson, "Bridging the Gap between Developmental Systems Theory and Evolutionary Developmental Biology," in *Bioessays*, No. 23 (2001), pp. 954–62 is useful.

CHAPTER 6. THE LOGIC OF DEVELOPMENT

The reader interested in a modern work on developmental biology can profitably consult the by now classic treatise by S. F. Gilbert, *Developmental Biology*, Sunderland, MA: Sinauer Associates, 1985, 2006, or the volume by L. Wolpert, R. Beddington, J. Brockes, T. Jessell, and E. Meyerowitz, *Principles of Development*, London: Current Biology, 1998, 2001; the latter devotes more space to the role of genic expression in development.

The by now century-old debate on the relation between the development of the individual and the evolution of the species was described and brilliantly commented on by S. J. Gould in *Ontogeny and Phylogeny*, Cambridge, MA: Belknap Press of Harvard University Press, 1977.

On the passage from a unicellular to a multicellular state, one should consult the brilliant essays by L. W. Buss, *The Evolution of Individuality*, Princeton: Princeton University Press, 1987, and by J. T. Bonner, *First Signals: The Evolution of Multicellular Development*, Princeton: Princeton University Press, 2001.

The idea that developmental processes should be interpreted according to their intrinsic logic, stage after stage, instead of in an "adultocentric" perspective, has been expressed by this book's author in chapter 2 of the book *The Development of Animal Form*, cited above, and by J. S. Robert, *Embryology, Epigenesis and Evolution: Taking Development Seriously*, Cambridge: Cambridge University Press, 2004.

Chapter 7. Paradigm Shifts

On the evolution of the concept of gene in the course of the twentieth century, one should consult the important collection edited by P. Beurton, R. Falk, and H.-J. Rheinberger, *The Concept of the Gene in Development and Evolution*, Cambridge: Cambridge University Press, 2000.

Chapter 8. Comparisons

The scientific literature on the concept of homology, and, more generally, on the comparative method, is extremely vast, but popular accounts are scarce. For a more detailed review, the volume edited by B. K. Hall, *Homology: The Hierarchical Basis of Comparative Biology*, San Diego and London: Academic Press, 1994, and the collection edited by G. R. Bock and G. Cardew, *Homology*, Chichester: Wiley, 1999, can be useful starting points.

Chapter 9. The Body's Syntax

On this topic, it is useful to consult chapter 8 of my book *The Development of Animal Form*, cited above.

Chapter 10. Competition or Cooperation?

An innovative interpretation of the embryonic leaflets (in vertebrates) has been presented in B. K. Hall, *The Neural Crest in Development and Evolution*, New York: Springer, 1999.

On miniaturized animals, one might consult the interesting critical review by J. Hanken and D. B. Wake, "Miniaturization of Body Size: Organismal Consequences and Evolutionary Significance," in *Annual Review of Ecology and Systematics*, No. 24 (1993), pp. 501–19.

A brief introduction to the biology of *Dictyostelium* can be found in the volume by Wolpert and others, *Principles of Development*, cited above.

Chapter 11. Making and Remaking

For a more technical exposition of the ideas mentioned in this chapter (double segmentation and the origin of appendages most particularly), the reader may usefully consult chapters 8 and 9 of my book *The Development of Animal Form*.

Chapter 12. Innovations without Plans

Differing points of view regarding the concept of module in its possible applications to developmental and evolutionary biology are presented in the volume edited by G. Schlosser and G. P. Wagner, *Modularity in Development and Evolution*, Chicago and London: University of Chicago Press, 2004.

Original ideas relating to the origin of evolutionary novelties can be found in the recent book by W. Arthur, *Biased Embryos and Evolution*, Cambridge: Cambridge University Press, 2004.

Index

Note: Page references in italics refer to figures.